Political Violence and Stability in the States of the Northern Persian Gulf

Daniel L. Byman

Jerrold D. Green

Prepared for the
Office of the Secretary of Defense

RAND
National Defense Research Institute

Political violence—terrorism and politically motivated killings or destruction intended to advance a political cause—has taken the lives of hundreds of U.S. soldiers and civilians in the Middle East in the 1980s and 1990s, and remains a serious threat for the coming decades. Its dangers go beyond lost lives: Political violence can create a climate of unrest in a critical region, leading once-stable countries such as Lebanon and Algeria to descend into an inferno of strife and civil war. In 1995 and 1996, terrorist attacks in Saudi Arabia killed 24 U.S. soldiers, and the possibility for further violence remains real. These terrorist attacks also raise a broader threat to the security of the U.S. regional presence and the stability of area regimes. In a worst-case scenario, terrorists also might act in conjunction with regional aggressors, helping them strike behind the lines of U.S. allies and impeding a U.S. military buildup.

This report assesses the threat of political violence in the northern Persian Gulf states of Saudi Arabia, Kuwait, Bahrain, and the United Arab Emirates. It examines general sources of discontent in the Gulf, common reasons for anti-regime politicization, potential triggers of violence, and the influence of foreign powers. The report then assesses those strategies that regimes in the area have used to interfere with political organization and to counter violence in general. The report concludes by noting implications of political violence for both the United States and its allies in the Gulf.

This assessment is intended to inform both policymakers and individuals concerned with Persian Gulf security. Policymakers can draw on the assessment in judging how to better protect U.S. forces and to understand the true level of threat to Gulf regime stability.

This research was conducted for the Office of the Assistant Secretary of Defense (Special Operations and Low Intensity Conflict/Policy Planning) within the Center for International Security and Defense Policy of RAND's National Defense Research Institute, a federally funded research and development center sponsored by the Office of the Secretary of Defense, the Joint Staff, the unified commands, and the defense agencies.

CONTENTS

FIGURES

TABLE

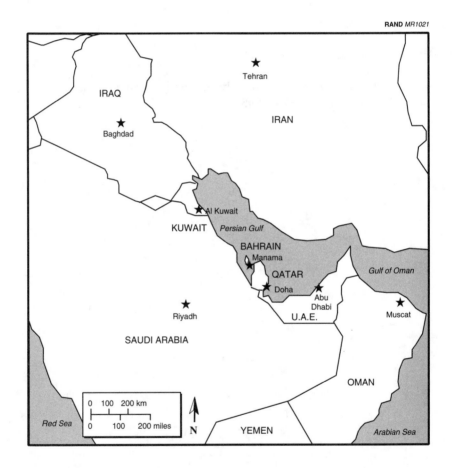

Persian Gulf Area

Political violence—terrorism and politically motivated killings or destruction intended to advance a political cause—has taken the lives of hundreds of U.S. soldiers and civilians in the Middle East in the 1980s and 1990s, and remains a serious threat for the coming decades. In 1995 and 1996, terrorist attacks in Saudi Arabia killed 24 U.S. soldiers, and the possibility for further violence remains real. Its dangers go beyond lost lives: Political violence can create a climate of unrest in a critical region, leading once-stable countries such as Lebanon and Algeria to descend into an inferno of strife and civil war. These terrorist attacks also raise a broader threat to the security of the U.S. regional presence and the stability of area regimes. In a worst-case scenario, terrorists also might act in conjunction with regional aggressors, helping them strike behind the lines of U.S. allies and impeding a U.S. military buildup.

Political violence in the northern Persian Gulf—Saudi Arabia, Bahrain, Kuwait, and the United Arab Emirates (UAE)—is particularly worrisome, because this region is critical for the United States' and the West's energy security and is threatened by Iran and Iraq. Although the 1995–1996 attacks did not diminish U.S. support for a strong regional presence, similar attacks elsewhere in the Middle East have led both Congress and the American people to question the desirability of U.S. overseas deployments.

To determine the extent of the problem, this report analyzes the threat of political violence in the Gulf, challenges to stability, and the effectiveness of countermeasures by Gulf governments. To do so, it looks at the scope of the threat, examining the broad grievances in the Gulf stemming from social, political, and economic problems. It then explores various politicizing factors that might make individuals turn against their governments; identifies potential "triggers," events that could lead to a sudden shift in popular attitudes; and assesses the influence of foreign powers on Gulf stability. To balance the potential for violence, it next explores the various tools that Gulf governments use to prevent violence, promote stability, and impede anti-regime political organization. Given the many components of violence and violence prevention, the report ends by discussing the difficulties that the United States and its Gulf allies will face in fighting political violence in the future.

A RANGE OF GRIEVANCES

The northern Gulf states face a wide array of problems that could lead to domestic instability and, eventually, to political violence. Demographic and economic problems are at the root of many grievances Gulf citizens express about their regimes. In the 1970s, the Gulf states—flush with billions in petrodollars—created expansive welfare states, providing their citizens with free education, health care, and other benefits. Any citizen receiving a college degree was guaranteed a high-paying government job. Gulf economies have since declined or stagnated, but population growth in the region has averaged almost 4 percent a year in the past two decades. As a result, the large youth populations of the Gulf expect high-paying, undemanding government jobs while regimes have fewer resources with which to satisfy them.

Adding to this structural problem are the rampant corruption and conspicuous consumption of the ruling families. Not surprisingly, the excesses of the ruling families generate considerable resentment as Gulf citizens are forced to tighten their belts. Unfortunately, the Gulf ruling families are taking few steps to liberalize their economies, improve education, reduce corruption, or otherwise increase the chances for sustained economic growth.

Political and social problems compound the resentment that stems from demographic and economic concerns. Citizens generally have little influence over decisionmaking and no way to ensure that government officials are held accountable for their actions in a region whose politics are dominated by ruling families and their close associates. A rapidly changing society also causes unrest. Many "traditional" citizens of the Gulf states, those who are suspicious of social change and seek to preserve the ways and mores of their ancestors, are upset by the perceived sexual promiscuity, drug use, and other modern evils that they fear are seeping into their societies.

The United States is a focal point for much of the resentment generated by the above problems. Many Gulf citizens believe that Washington exercises extensive control over the Gulf regimes and opposes reform for its own selfish purposes. U.S. support of Israel and, to a lesser extent, U.S. hostility toward Iraq and Iran also generate resentment. Perhaps more important, many Gulf citizens— especially in Saudi Arabia—are angered by the large U.S. presence in the region. Radical Saudi Islamists oppose the stationing of non-Muslim forces on Saudi soil. Saudis generally question the cost of maintaining the U.S. presence in the Kingdom and of buying U.S. arms at a time when the Kingdom's own economy is stagnating.

DIFFERENT VULNERABILITIES TO POLITICAL VIOLENCE, BY STATE

To understand the threat of political violence in the northern Gulf, we need to look at each country in the region and the particular problems it faces.

Bahrain faces the gravest problems of any Gulf state, although its unrest has not been accompanied by anti–U.S. activities. Unlike the other Gulf states, Bahrain has almost no oil reserves with which to co-opt its population. That is, the Al Khalifa ruling family can offer few high-paying government positions or lucrative contracts to win the goodwill of important citizens and groups. To preserve peace, the ruling family relies heavily on financial support from other Gulf states and on its efficient, but ruthless, security services. Corruption is widespread in Bahrain: Getting a contract or doing business with the government often requires giving ruling-family members a cut of total profits. Moreover, the ruling family adheres to the Sunni sect of Islam, while over 70 percent of Bahrainis are Shi'a Muslims. The regime makes little effort to give Shi'as significant political influence, and the Shi'a usually face rampant discrimination and poor economic prospects. In 1994, 1995, and 1996, demonstrations, arson, stone-throwing, and other limited forms of violence regularly occurred in Bahrain, and sporadic unrest remains a problem. The regime remains in control, but opposition forces appear to have the sympathy of many Bahrainis. Even some Sunni elites hesitate to support the royal family, and the regime's ability to win over Bahrain's majority Shi'a population—particularly its poorer members—is questionable.

Facing fewer problems than does Bahrain, Saudi Arabia may encounter unrest from ultraconservative Sunnis. The Al Saud have generally kept the peace by lavishing financial support on tribal and religious leaders while suppressing Shi'a agitation or deflecting it through co-optation and financial support. Yet, although the Kingdom's oil reserves are substantial, its growing population is straining the regime's ability to buy off dissenters. Islamist sentiment is particularly strong in Saudi Arabia, and most opposition is expressed in terms of the regime's failure to adhere to religious tenets or practices. Many anti-regime militants also blame the United States for the Kingdom's problems and consider U.S. forces a legitimate target for expressing their discontent with the regime. Data on Saudi Arabia, its dissenting groups, public opinion, Shi'a community material conditions, government spending, and other important concerns are scarce, however, and the political opposition may be more active and more organized than we recognize.

Kuwait and the UAE face few political or economic problems compared with Bahrain and Saudi Arabia. In both countries, the high levels of wealth and the government's willingness to share liberally with all segments of society have kept most citizens reasonably content. The UAE has virtually no political opposition, and its President, Shaykh Zayid, is generally popular. The Al Sabah in Kuwait allow Kuwaitis a limited voice in decisionmaking: Kuwaitis have an elected National Assembly with some limited political powers, and the ruling family makes an effort to win over critics. These measures contribute to the ruling family's popularity. The constant Iraqi threat to Kuwait also unites Kuwaitis, reducing the scope of anti-regime activity. Both countries' security services have proven skilled at countering foreign-backed opposition. To date, neither has had to contain a major anti-government movement.

POLITICIZATION AND ORGANIZATION IN THE GULF

Although political alienation, economic stagnation, and unwanted social change are common problems in the Gulf, they seldom lead to violence. Discontent may be widespread, but few Gulf residents actively oppose their regimes, and even fewer use violence to do so. However, a range of problems could politicize small groups in a violent manner, leading them to use violence rather than peaceful means to express their grievances and promote instability. These problems include dissatisfaction with the political system, a desire to defend their traditional ways of life, and the glorification of violence.

Political alienation is widespread and significant in the Gulf, particularly in Bahrain and Saudi Arabia. Gulf political systems are exclusive—the ruling families monopolize decisionmaking—and, despite petitions and the formation of token national assemblies, many Gulf citizens correctly believe they have little or no influence on decisionmaking. If otherwise-moderate individuals conclude that peaceful political action will yield nothing, they may resort to violence.

Small groups of individuals—but sufficient numbers to conduct political violence—also may turn to violence in response to regime crackdowns on dissent. In Bahrain and Saudi Arabia, the ruling families regularly crush political opposition, including that of fairly moderate groups, which do not seek to use violence to advance their agendas. Individuals who are threatened with exile, jail, or worse might go underground to survive.

The rapid transformation of society is another common problem in the Gulf. Terrorists in other parts of the world—particularly those with religious or ethnic agendas—often take up arms because they see traditional ways of life as being under assault. The influx of oil wealth, and sudden exposure to the West and the broader Muslim world, has dramatically changed the Gulf. The traditional ways of life in the Gulf, present a mere 40 years ago, are now largely gone. Their end has promoted bitter responses from those who most cherished the old order.

Politicization also can occur when violence is glorified in the Gulf. At an intellectual level, and often in practice, both Arab nationalism and political Islam have embraced the use of violence, sometimes justifying it as a necessary instrument of politics. This support is reinforced at a popular level: Fighters in Afghanistan, Lebanon, Bosnia, and elsewhere are lauded as heroes, a glorification that legitimates violence and increases the status of those who use it.

Despite this potential for unrest, Gulf opposition movements are seldom organized, and Gulf regimes use a variety of measures to hinder anti-government and anti–U.S. organizations. The most-effective form of political organization in the Gulf occurs through religious channels. Radicals—those who seek to use violence to advance their political agendas—often exploit the network of mosques and prayer groups, although area regimes carefully monitor most religious activity, particularly potentially oppositional activity.

This lack of organization limits more-dangerous forms of political opposition, such as a systemic terrorist campaign or an insurgency. When individuals cannot organize, it is far more difficult for them to train and gather intelligence, both of which are necessary to attack well-guarded institutions. Creating a climate of dissent is difficult, because regime security services will suppress local groups quickly, and it is even more difficult for opposition groups to work systematically with foreign militaries.

Because of this lack of organization, political opposition and violence often occur spontaneously. Individuals frequently act with little planning, without well-defined objectives, and without coordinating their actions with those of other like-minded individuals. Such actions still can lead to the deaths of U.S. and Gulf citizens, but they are not likely to topple area regimes or pose an immediate threat to U.S. operations.

THE ROLE OF FOREIGN POWERS

Foreign support for a regime's violent opponents and/or terrorists can make those groups far more effective and deadly. Iran, Iraq, and transnational Islamic movements could conceivably support local activists in the Gulf against their regimes and the United States.

Iran has actively supported political violence as part of its foreign policy. Iran tried to create local proxies to carry out its wishes and spread its revolutionary credo. In addition, the Islamic Republic has used political violence to assassinate opponents, harass rival governments, and demonstrate support for worldwide Islamic causes. Although Iran has not directly supported violence in the Gulf in recent years (the jury is still out, however, on who was responsible for the 1996 Khobar Towers attack), it has tried to create local proxies that are armed and capable of using violence at Tehran's behest.

Iran is not the only foreign actor that might support violence in the Gulf. Iraq has regularly employed political violence, assassinating regime opponents throughout the Arab world and even attempting to kill former President George Bush in 1993. Given Saddam Husayn's ongoing hostility toward the United States and its Gulf allies, Iraq is likely to use political violence to advance its agenda. The Lebanese Hezbollah also has worked with Gulf citizens, inculcating Islamist teachings and perhaps training them in terrorist tactics and arming them as well.

Foreign powers greatly influence the potential for the growth of political violence and the efficacy of strategies designed to contain it. Foreign powers can contribute to a general atmosphere of discontent, using their media and ties to local elites to attack the legitimacy of a regime and highlight grievances against it. Outside powers also can provide a model for political action, inspiring Gulf residents with their example and their message. Even more important, outside powers can help opposition groups organize and arm, providing them with intelligence and weapons, and a haven in which to organize.

STRATEGIES TO FIGHT DISSENT

Gulf regimes maintain power and preserve stability with the deft application of "sticks" and "carrots." Strong security services interfere with opposition organizations, convincing potential leaders and their followers that the price of dissent is too steep by imprisoning or harassing oppositionists. When heavier-handed tactics fail, Gulf regimes have proven skilled at gentler tactics as well. The ruling families are masters of co-opting critics by providing them positions in government, financial incentives, and other rewards for cooperating. Gulf governments also divide potential oppositionists by playing on tribal, regional, religious, and class distinctions and creating rifts in the populace at large and among elites to keep all opposition weak. At the same time, ruling families often champion opposition causes, nominally supporting a wide range of agendas in order to undercut and divide the opposition and bolster support for the regime. When necessary, Gulf governments have allowed citizens limited voices in decisionmaking, which can defuse political alienation. Finally, Gulf governments employ conciliatory foreign policies, declaring their friendship as a way to prevent foreign powers from stirring up unrest.

Several tools the Gulf states use to combat political violence and anti-regime organizations may sometimes actually worsen these problems. For example, Bahrain's heavy reliance on its security services increases political alienation and may be turning reformers into revolutionaries. Efforts to appease foreign aggressors may lead Gulf regimes to tolerate the spread of potentially subversive ideologies.

IMPLICATIONS FOR THE UNITED STATES AND ITS ALLIES

The Gulf states face little organized threat today and appear secure for the foreseeable future. They have weathered the successive storms of Arab nationalism and Islamic radicalism, and emerged with a strong grip on power. Although problems abound, Gulf leaders have shown themselves skilled at repressing and placating their citizens and preserving their hold on power.

This picture is not completely rosy. Gulf regimes will never be able to prevent all forms of political violence from occurring. Violence by small groups will remain a constant problem. Unorganized groups, while far less lethal or politically effective than organized movements, still may attack U.S. or regime personnel and facilities.

Many of the problems facing the Gulf states are not easily rectified. Some factors, such as opposition to social change, are simply beyond the control of almost all governments. Alleviating other sources of resentment, such as economic problems and political exclusion, will require drastic changes in the way the ruling families govern. Completely eliminating foreign-backed political violence is also beyond the control of the Gulf states. Effective solutions to Gulf political violence will have to take these factors into account. Moreover, solutions must be altered to fit the particular needs of each state; for example, what works for Bahrain may fail in Saudi Arabia, or vice versa.

Several measures might reduce the problem of political violence, helping decrease overall popular animosity and increasing the ability of the United States and its allies to fight radicals. However, the record so far suggests that some of these measures will be difficult to implement and that their impact is likely to be limited. The measures include the following:

- **Instituting Political and Economic Reforms.** Political and economic liberalization could reduce many grievances of Gulf citizens. Increasing popular participation in decisionmaking and efforts to open up the economy by reducing government oversight could offset the widespread anger at corruption and the ruling family's tight grip on power. However, Gulf governments lack both the inclination and the resources to engage in political and economic reforms that could alleviate much of the discontent currently found in the Gulf. Even if they nevertheless undertook reform, belt-tightening would probably increase dissent in the short term. In addition, political liberalization could confer more freedom on individuals and groups that strongly oppose the U.S. regional presence.

- **Confronting Foreign Interference.** Gulf governments are skilled at countering efforts by Tehran and Baghdad to provoke unrest. To further offset Iranian (and possibly future Iraqi) meddling, the United States could threaten to respond even more forcefully than it already does to acts of terrorism and subversion in the Gulf. Iran and Iraq have demonstrated a healthy respect for U.S. capabilities and probably would hesitate before striking. The U.S. ability to coerce both Iran and Iraq is limited, however. Moreover, the United States' Gulf allies often prefer accommodation over confrontation, and military strikes would complicate U.S. relations with allies in the region.

- **Reducing the U.S. Presence.** Decreasing the overall size of the U.S. regional presence in the region is a useful measure for fighting political violence. Fewer U.S. troops in the region would decrease both the political and material strain on Gulf allies while placing fewer U.S. soldiers in harm's way. Of course, the remaining troops would face a threat from regional radicals, and the Gulf states also will still be criticized by those opposed to *any* U.S. presence. Basing outside the Gulf region and instituting new organizational and operational approaches might offset the military disadvantages inherent in drawing down the U.S. presence.

- **Encouraging a Greater European Role.** A greater European role could reduce the number of U.S. troops exposed to violence and offset any potential criticism in the United States over its presence in the Gulf. But such participation may require a political price: European governments have proven more open to working with Iran and Iraq than has the United States. European allies have so far evinced little desire to play a major role in Gulf security, and their militaries lack the necessary assets to contribute substantially to Gulf defense. However, several European countries are currently considering improving power projection and preparing for missions outside of Europe, which could make this a more realistic option in the future.

- **Strengthening the U.S.–Gulf Partnership.** Greater U.S.–Gulf state cooperation would help combat violence, but several obstacles hinder improved ties. Gulf states are angered at the legalistic approach the United States employs toward terrorism. To Gulf regimes, Washington demands too high a level of evidence before acting or retaliating. Media leaks that reflect poorly on Gulf leaders further anger them. Finally, both the United States and the Gulf are reluctant to share sensitive intelligence. As a result, cooperation against radicals is often limited.

- **Increasing Military-to-Military Ties.** Increasing contacts between military officers of Gulf states and those of the United States might improve relations and provide additional intelligence. These contacts would strengthen the overall U.S.–Gulf partnership, although they would do little to win the goodwill of those in the Gulf most hostile to the West, particularly Islamists.

The above measures can help reduce political violence, but they will not stop it altogether. Given the likelihood that low levels of political violence will continue, maintaining the high level of attention to force protection is mandatory. Although the situation in the Gulf currently appears peaceful relative to other parts of the Middle East and developing world, eliminating all political violence may prove impossible. Moreover, as the August 1998 attacks in Kenya and Tanzania suggest, the Gulf is not the only theater for political violence. Violence emanating from events in the Gulf is spreading throughout the world, and success in force protection or anti-terrorism in the Gulf may goad terrorists to strike at less-defended targets outside the Gulf region. Given such limits, improving the personal protection of U.S. soldiers in the Gulf and throughout the world is crucial. Not only will such protection save the lives of U.S. personnel, it also will reinforce existing popular and congressional support for the regional U.S. presence.

ACKNOWLEDGMENTS

Many individuals aided in the research and preparation of this report. Bruce Hoffman and F. Gregory Gause III provided superb, detailed reviews of earlier drafts of this work, strengthening it considerably. Among our RAND colleagues, we would like to thank Jeff Isaacson for his many insights, and Donna Boykin, Cyndi Lockwood, and Suzanne Newton for their invaluable administrative assistance. Marian Branch did an excellent job of editing this study. Gail Kouril and Jennifer Ingersoll-Casey made the research on this project far easier and more fruitful. Omid Fattahi conducted useful research on terrorist incidents in the Gulf. Outside RAND, the authors would like to thank Michael Eisenstadt, Hillary Mann, Kenneth Pollack, and Judith Yaphe for their assistance in finding data for various chapters of the report.

The authors would like to thank U.S. officials in the military services, Department of Defense, Department of State, Central Intelligence Agency, and Federal Bureau of Investigation for their assistance in this project and willingness to provide insights into political events in the Persian Gulf. Finally, the authors thank officials and experts in the Gulf itself for their time and efforts on our behalf.

Chapter One

INTRODUCTION

POLITICAL VIOLENCE AS A THREAT TO THE U.S. PRESENCE IN THE GULF

In times of peace, terrorists and other practitioners of political violence—not conventional military forces—may pose the greatest threat to the lives of U.S. soldiers and the security of U.S. allies.[1] A soldier entering the Army in 1977 and retiring today would have been more likely to die from a terrorist attack than be killed in combat. As Table 1.1 suggests, political violence has been a constant problem for the past 25 years, occurring for a variety of reasons and involving both local Gulf groups and foreign sponsors. Many attacks occurred at some distance from the Gulf. Given this long-standing danger from terrorism and other deliberate, politically motivated killings or destruction intended to advance a political cause, the problem requires increased scrutiny today.

---

[1]The definition of *terrorism* is highly controversial. See Hoffman (1998), pp. 13–44, for a discussion. This chapter follows Hoffman's definition (p. 15), and uses the term terrorism to mean "violence—or, equally important, the threat of violence—used and directed in pursuit of, or in service of, a political aim." Hoffman, however, also notes that terrorism is "a planned, calculated, and indeed systematic act." For purposes of this study, *political violence* is used as a broad term that encompasses terrorism *and* includes violent political acts that are not planned or systematic. For example, the Hezbollah destruction of the U.S. Marine barracks in Lebanon qualifies as both terrorism and political violence; a Bahraini rioter throwing stones at a policeman would be engaged in political violence, but not terrorism. The line between a terrorist and a radical is also blurry. Many radicals attack civilian targets in an effort to achieve their political goals—activities that make them like terrorists in essence. In this study, *radicals* are those who regularly use violence to further their political agendas.

Table 1.1

The Evolution of Politically Directed Violence in the Middle East, 1975–1996

Year	Month/Day	Description
1975	March 25	King Faisal of Saudi Arabia is assassinated by a kinsman and is succeeded by his brother Khalid
1978	July 31	The Iraqi Embassy in Paris is seized by an Al-Fatah terrorist (who later surrendered and was wounded by Iraqi guards)
1979	November 4	Iranian students attack and occupy the U.S. Embassy in Tehran and hold 52 Americans hostage for 444 days
1979	November 20	A group of Sunni Muslims, claiming that the *Mahdi* (returned prophet who offers deliverance) was among them, seize the Grand Mosque in Mecca for 15 days before being driven out by Saudi security forces; up to 200 of the militants are killed
1980	April 30	Iranian Arabs seize the Iranian Embassy in London, taking 26 hostages and subsequently killing 2 before the building is retaken by British security forces
1981	June 27	The prime minister of Iran and 70 others, including Ayatollah Beheshti, are killed in the bombing of the Majlis building in Teheran
1981	December	A plot to overthrow the Bahraini government by the Iranian-backed Islamic Front for the Liberation of Bahrain is foiled
1983	April 18	Iranian-backed radicals blow up U.S. Embassy in Beirut
1983	October 23	Suicide-bombing of U.S. and French troops in Lebanon
1983	December 12	Da'wa radicals explode car bombs in Kuwait
1984	December 4	Four Islamic Jihad terrorists hijack a Kuwaiti airliner bound for Pakistan and order it to fly to Tehran; 2 are killed while 2 others, including a U.S. businessman, are tortured; Iranian troops storm the aircraft and free the hostages on December 9
1985	May 25	An assassination attempt is made on Kuwaiti Emir Shaykh al-Sabah; 3 are killed and 15 wounded
1989	June	A religious ruling calling for the death of British author Salman Rushdie is declared in Iran following the publication of his new book *The Satanic Verses*
1989	July 13	Abdel Rahman Qassemlou, leader of the Kurdish Democratic Party of Iran, is assassinated by Iranian agents in Austria

Table 1.1—continued

Year	Month/Day	Description
1991	August 6	Former Iranian Prime Minister Shahpour Bakhtiar and an aide are murdered in Paris
1992	March 17	A bomb explodes at the Israeli Embassy in Buenos Aires, killing 29 and injuring 242; Iranian involvement is suspected
1992	April	The Iranian Mujahedin-e-Khalq carries out simultaneous attacks on 13 Iranian embassies in North America, Europe, and the Pacific Rim
1992	September 17	Four Iranian Kurdish dissidents, including the leader of the Kurdish Democratic Party of Iran, are assassinated at a Greek restaurant in Berlin; in 1997, a German court charges high Iranian government officials, including President Rafsanjani and Supreme Leader Khamenei, with approving the attack
1993	March 17	Explosion at Israeli Embassy in Beirut, probably by Hezbollah-affiliated radicals
1993	April 15	An Iraqi plot to assassinate former U.S. President Bush while on a visit to Kuwait is uncovered; the U.S. launches a retaliatory military strike against Iraq on June 26
1994	March 11	Iran is suspected of involvement in an abortive attack on the Israeli Embassy in Bangkok
1994	June 20	A bomb explodes at the Imam Reza mausoleum in Meshhed, Iran, during ceremonies commemorating the death of Imam Hussein; the Mujahedin-e-Khalq claims responsibility
1994	July 18	A car bomb explodes outside the Argentine-Israeli Mutual Association in Buenos Aires, killing 100 and wounding 200; Iranian involvement is again suspected
1995	November 11	A car bomb badly damages the headquarters of the Office of Program Manager/Saudi Arabian National Guard, a military training mission, in Riyadh; 7 people, including 5 Americans, are killed and 42 others wounded; 3 groups, including the Islamic Movement for Change, claim responsibility for the bombing
1996	June 25	A car bomb explodes outside the U.S. military's Khobar Towers housing facility in Dhahran, Saudi Arabia, killing 19 U.S. military personnel and wounding 515; the incident is still under investigation

This study argues that the threat of political violence can be reduced but is not likely to be eliminated. Even if progress is made on reducing overall dissatisfaction with Gulf governments and U.S. forces, threats will remain to U.S. personnel, to domestic support for the U.S. presence, and to allied stability. Because of the inherent limits to the U.S. ability to end political violence once and for all, a robust force-protection policy is necessary—particularly in the Gulf, because of the unorganized nature of any likely violence. Therefore, the current U.S. emphasis on force protection is sound, because it recognizes the imperfect nature of any solutions to the fundamental causes of political violence.

However, the danger of political violence extends beyond the tragedy of lost U.S. lives and force protection issues. In the Middle East, political violence has destabilized friendly governments, and could continue to do so, to the benefit of foreign aggressors. Fearing violence if they cooperate with Washington, regional allies might be intimidated into denying the United States access to the region if a crisis arises.

The region's criticality means that political violence in the northern Gulf states—Saudi Arabia, Bahrain, Kuwait, and the United Arab Emirates (UAE)—poses a particular challenge to U.S. interests.[2] Saudi Arabia, Kuwait, and the UAE have huge oil and gas reserves, which are vital to the United States and central to the world's energy security for the foreseeable future. In the past, Iran and Iraq threatened the security of these producers and the United States' and its Western allies' access to oil. In addition to direct military threats to the northern Gulf states, Iran and Iraq have used the Gulf states as a surrogate battlefield in their struggles against each other.

U.S. forces in the Gulf area carry out a variety of missions. Most of the missions are to deter Iran and Iraq or, should deterrence fail, to defeat aggression. The United States conducts Operation Southern Watch—the enforcement of a "no-fly" zone over southern Iraq—primarily from bases in Saudi Arabia. About 5,000–6,000 U.S. troops are stationed outside of Riyadh or at nearby Al Kharj, as part of the Joint Task Force Southwest Asia (JTF-SWA), which is responsible for

---

[2]This study does not address the question of political violence in Qatar and Oman; our focal point was exclusively the states of the northern Gulf.

Southern Watch. The Fifth Fleet, with a staff of fewer than 1,000 personnel, projects U.S. power throughout the Gulf and is headquartered in Bahrain. Although the United States has no formal forward presence in Kuwait, about 1,500 U.S. soldiers are there four times a year for 3-month-long exercises, providing, in effect, an informal year-round forward presence. The United States does not station troops in the UAE, but the port in Dubai has become the most-visited U.S. Navy port of call outside the continental United States. In times of crisis, this troop presence is bolstered and can total more than 30,000.

Going beyond troop strength alone, the true U.S. presence in the region includes a variety of security and access agreements, agreements that are essential if the United States is to defend the Gulf against an aggressor. With the exception of Saudi Arabia, all the Gulf states have signed defense cooperation agreements with the United States. Qatar and Kuwait agreed to house prepositioned equipment for an Army brigade. The United States has also arranged for several Air Expeditionary Force (AEF) wings to deploy to locations in Kuwait, Bahrain, and Qatar. The prepositioning and regular exercises concomitant with the U.S. presence supply the infrastructure necessary for the rapid insertion of ground forces.

But a serious vulnerability accompanies this presence: a vast and diverse range of potential military targets for anti–U.S. radicals to strike. Large numbers of U.S. personnel in all the states can be attacked by terrorists, as can U.S. facilities, as the 1995 and 1996 terrorist killings of U.S. military personnel in two attacks in Saudi Arabia attest.

OTHER CONSEQUENCES OF POLITICAL VIOLENCE

In addition to the tragedy of lost U.S. lives, political violence in the Gulf can lead to three broad problems for the United States:

- Growing pressure at home to withdraw U.S. forces or not commit them in the event of a conflict

- A climate of unrest that leads Gulf leaders to deny U.S. forces access to the region, thus leaving the region vulnerable to an attack from Iran or Iraq

- Operational problems if radicals act in conjunction with aggressor states, such as threats to access and behind-the-lines enemy assaults.

Violence by radicals can increase domestic pressure in the United States, resulting in a force drawdown. The deaths of U.S. servicemen or citizens from political violence can lead the American public to question the purpose of the U.S. presence while making policymakers hesitant to send more soldiers into harm's way. In Lebanon, for example, the death of 241 Marines led to the end of the U.S. deployment there. In Somalia, the deaths of 18 U.S. soldiers led the United States to remove its forces from UN operations there. Radical groups of all sorts have correctly concluded that the United States is reluctant to risk casualties and that political violence can force the United States to reduce its presence. For now, support for the regional U.S. presence appears solid. However, pressures resulting from future attacks in the Gulf could force an administration to withdraw troops despite threats to regional security and U.S. interests.

Even if the United States is resolved to maintain its regional presence, political violence can cause Gulf states to deny the United States the access required to do so. Gulf governments in general favor a continued U.S. presence, but political violence directed against U.S. forces might cause them to choose between continuing internal instability and the risk of external aggression if U.S. forces depart. A climate of unrest in Saudi Arabia could make the Al Saud hesitant to accept a large American presence or any augmentation. For example, radicals could conduct strikes in Saudi Arabia to press Riyadh to reduce the U.S. presence in the Kingdom, a presence necessary to enforce the no-fly zone over Iraq. Disagreements over how to respond to the threat of political violence in the wake of the 1996 Khobar Towers bombing, in which 19 U.S. military personnel died, have increased U.S.–Saudi tensions. The United States has publicly criticized the Saudis for not cooperating fully with investigation of the bombing. Should a crisis occur involving external aggression, this reluctance to tolerate a large U.S. presence may leave the Kingdom vulnerable militarily.

A climate of unrest also is dangerous in that it threatens the stability of allied regimes. Sustained violence could create the impression that the Gulf regimes are unable to maintain public order, thereby discrediting the police and security services, and leading activists of all sorts—individuals who are highly committed to political agendas, both radical and moderate—to challenge the regime. In addition, the success of violence in forcing a regime to respond may discredit moderates among the population and in the government. A widespread campaign of violence could even lead to regime paralysis or a new government.

Future violence also could strain relations between the United States and its allies in Europe and Asia. U.S. allies contribute little to the defense of the region, even though they depend heavily on Gulf oil. Should U.S. casualties grow, this lack of allied support could lead to U.S. resentment of that parsimony.

Tremendous problems also could occur operationally if terrorists act in conjunction with regional aggressors. Terrorist attacks on key ports or airfields could prevent or delay a deployment to that area, causing valuable time to be lost. The possibility of terrorist strikes against U.S. personnel also will make mobility within the theater more difficult, because force protection will become a leading operational concern.

Even the *possibility* of political violence could impede U.S. operations. Already, the United States devotes a tremendous amount of resources and planning to force protection. The United States has redeployed its forces in Saudi Arabia, minimized leave, and expanded passive defenses. If terrorists and other Gulf opposition figures appear likely to act on behalf of foreign governments, the United States may be constrained to deploy its forces to bases in more-remote areas, delay the deployment to ensure base security, or otherwise take steps that will hamper effectiveness against foreign armies.

This range of threats varies, in part, by the skill and size of the terrorist group. Small, fairly unorganized groups—the most common type in the Gulf—will find it far harder to strike well-guarded targets and will not be able to coordinate their activities with those of foreign militaries. Unorganized groups will also find it harder to sustain a campaign of terror that could destabilize the region.

COUNTERING POLITICAL VIOLENCE

Like any complex phenomenon, political violence can be fought in different ways as different facets of the problem are confronted. To help policymakers understand and counter the threat of political violence, this report explores the following components of the political-violence problem:

- Broad political, social, and economic grievances that create a climate of unrest and dissatisfaction in the Gulf and factors that make individuals dissatisfied both with their regimes and with the U.S. regional presence (Chapter Two)

- Politicizing factors that lead individuals to become actively opposed to their regimes and to the U.S. presence in the Gulf, as well as the disorganization of political action in the Gulf (Chapter Three)

- Triggering factors that quickly transform opponents' latent anger into active resistance to Gulf regimes (Chapter Three)

- The role of foreign powers in influencing political violence in the northern Gulf, and the specific goals and doctrines of the Gulf states' neighbors, Iran and Iraq (Chapter Four).

Using the information on these components, we then offer answers to the following questions: How do Gulf regimes try to counter political unrest and to stop anti–U.S. violence (Chapter Five)? What are the future dangers the Gulf and the United States may face and the complications inherent in potential solutions (Chapter Six)?

Much of the information in one chapter may be revisited in another chapter, but from a different perspective. For example, a social grievance in Chapter Two may, in a specific context or country, become a politicizing factor discussed in Chapter Three, as well as a factor that, under more-extreme conditions, leads to active/violent opposition to a regime. Although the revisited information is usually the most important for explaining the problem of political violence, this report also seeks to present the broad array of potential problems that could surface as more-specific grievances in the future or that foreign powers could exploit. This approach is intended to help readers recognize the interrelationship between general problems, such as unemployment, and specific political problems, such as a political opposition movement's call for the regime to reduce foreign labor. This structure also helps illustrate the overall context for particular violent acts, enabling policymakers and students of the region to understand how to better prevent, disrupt, or defend against political violence.

This report seeks to inform general questions about the level of threat in the Gulf, both to U.S. forces directly and to allied regime stability, and the best means of reducing that threat. Those concerned with force protection in particular will benefit, as it provides an assessment of the likelihood of continued violence and the level of sophistication of radical groups. The study also will be of interest to those involved in policymaking regarding the Gulf, because it assesses Gulf stability and potential threats to that stability. Finally, this study should be of interest to students of political violence, because it examines the threat in a vital but less-studied region and explores regime strategies for countering violence. For all readers, this study should help identify problems that can be solved, those that can be reduced, and those that are likely to continue in spite of the best U.S. and Gulf regime efforts.

The authors consulted Gulf leaders, U.S. government officials, defense strategists, and academic experts for their insights into the questions explored in this report. When appropriate, the authors supplemented those interviews with information from secondary sources and government data.

Chapter Two

THE ATMOSPHERE OF DISCONTENT IN THE GULF

Political violence does not occur in a vacuum. The Gulf states offer an array of "growth media" for such violence, from explosive population growth to disruptive social change. To identify the nature of unrest and the potential for political violence in the northern Gulf states of Saudi Arabia, Kuwait, Bahrain, and the United Arab Emirates (UAE), this chapter explores the broad social, political, and economic grievances that have led to widespread popular dissatisfaction in the Gulf.

The United States is both a target of this discontent and a source of it. Although most of the dissatisfaction felt by Gulf citizens is against their regimes or unwanted social changes, many oppose the large U.S. presence in the region and are highly critical of U.S. policies. This sentiment is particularly strong among those most likely to use violence. As a result, radicals often criticize or attack the United States because of discontent with their domestic political systems.

THE POTENTIAL FOR UNREST IN THE GULF

This section explores the overall level of grievances held by Gulf citizens. In particular, it examines a gap in expectations, a lack of freedom or accountability, general economic problems, foreign-inspired violence, destabilizing social change, and dependence on the United States. Where appropriate, particular problems for selected Gulf states are noted. At the end of this section, we summarize these issues in graphic format for easy comparison.

In the Gulf, there is a gap between popular expectations of government and the regimes' ability to provide for their citizens—and the gap is growing, creating widespread unhappiness. Specific problems include demographic changes, corruption, a lack of government accountability, profligate royal-family spending, wealth disparities, poor economic performance, a lack of public participation in decisionmaking, and uncontrolled social change. All these grievances can lead to dissent and, eventually, to violence. Unfortunately, Gulf governments have taken few steps to carry out economic and political reform.[1]

Economic and Demographic Problems

The Gulf states—Bahrain and Saudi Arabia, in particular—suffer from a combination of high expectations, rapid population growth, and falling oil revenues. In the 1970s, the rapid influx of oil wealth led all the Gulf regimes to create extensive welfare systems. Their states provided free health care, education, and other services to all citizens. Moreover, citizens with advanced degrees were entitled to lucrative government jobs.

In the early 1980s, the price of oil fell dramatically and, although rising somewhat in early 1999, has not recovered.[2] At the same time, rapid population growth—some Gulf states have averaged gains of almost 4 percent annually in the past two decades—has created a large and restive youth population.

The rapid population upsurge indicated in Figures 2.1 and 2.2 is destabilizing for at least two reasons. First, rapid population growth

---

[1]This study does not examine the grievances and organization of the large expatriate Palestinian, Iranian, and other communities in the Gulf, except as they affect local Gulf citizens. Although, in theory, these groups could pose a threat to Gulf regimes, they are monitored vigilantly by Gulf security services. Many expatriate Palestinians and Iranians work in the Gulf; however, they have posed little threat to regime stability, even though many may be sympathetic to radical causes. Suspicions of political activity, let alone anti-regime violence, usually lead to immediate deportation of both the individuals and all associated with them. In some Gulf states, foreigners also receive free or cheap social services and pay no taxes (Rugh, 1997).

[2]In 1996, an estimated 65 percent of Bahrain's total revenues came from oil and gas. The figures for the UAE, Saudi Arabia, and Kuwait are 84, 73, and 73 percent, respectively (Sick, 1997, p. 17).

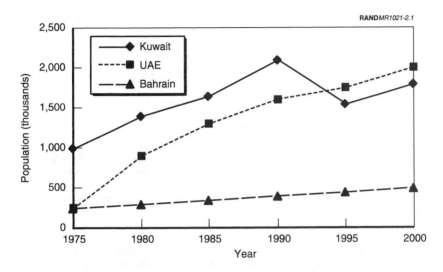

SOURCES: 1976, 1983/1984, and 1993 *Statistic Yearbooks; 1986 Demographic Yearbook; CIA Handbook of International Economic Statistics, 1996; CIA World Factbook, 1996;* Cordesman, 1997a, p. 10. The figures for the year 2000 are estimates. All figures on Gulf populations are highly speculative. Official population numbers include expatriate laborers, who do not enjoy the benefits of Gulf citizenship. Moreover, their numbers often fluctuate sharply in response to local economic conditions. These caveats aside, this figure accurately reflects, in general, the huge population growth in the Gulf. The Gulf as an area has had one of the highest population growth rates in the world in the past few decades. The one exception to this is the population of Kuwait from 1990 to 1995, which shrank in total numbers including expatriates, but grew in Kuwaiti citizens.

Figure 2.1—Population Growth in Kuwait, Bahrain, and the UAE, 1975–2000

generates tremendous economic pressure, "running just to stay in place": Simply to retain the same levels of wealth on an individual basis, Gulf economies must grow rapidly. Second, rapid growth exerts pressure on governments to expand education, medical care, and social services at breakneck speed—a pace that can lead to bottlenecks and inefficiencies even when governments have vast wealth. When government revenues are stagnant or declining—as they are in the Gulf today—regimes cannot satisfy these same pressures that rapid population growth creates.

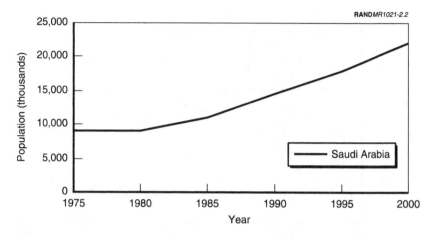

SOURCES: 1976, 1983/1984, and 1993 *Statistic Yearbooks; 1986 Demographic Yearbook; CIA Handbook of International Economic Statistics, 1996; CIA World Factbook, 1996.* The figures for the year 2000 are estimates. The same caveats noted in Figure 2.1 also apply to Saudia Arabia. Moreover, Saudi Arabia's population figures also are somewhat skewed, because the regime exaggerates its total population in order to strengthen its claims to being the preeminent power on the Arabian peninsula.

Figure 2.2—Population Growth in Saudi Arabia, 1975–2000

The Gulf's staggering population growth is particularly destabilizing because of the disproportionately large youth population that has resulted (Figure 2.3). And these young people generally have higher expectations than their elders. Most Gulf residents under the age of 30—easily more than two-thirds of the population—grew up accustomed to a high standard of living. They expect excellent health care, housing, and other services that their parents never knew as children. Furthermore, many received higher degrees, increasing their ostensible qualifications for high-status, high-paying jobs. Thus, expectations from government of the ever-growing numbers of these youths are escalating at a time when government revenues are diminishing. These younger citizens tend to favor radical causes while otherwise engaging in antiestablishment activities. Moreover, in contrast to their parents and grandparents, these youths have grown up in countries whose governments promised cradle-to-grave services. They bitterly resent cutbacks in benefits they view as entitlements—virtual birthrights.

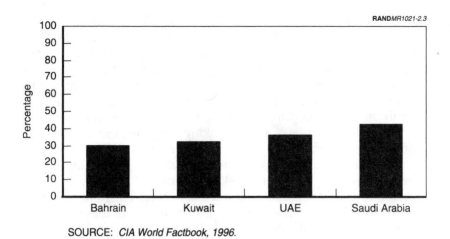

SOURCE: *CIA World Factbook, 1996.*

Figure 2.3—Percentage of Population Aged 0–14 Years

Some Gulf states lack the wherewithal to satisfy their burgeoning populations. The per-capita incomes of many Gulf residents plummeted after the price of oil began falling in the early 1980s. From 1984 to 1994, real per-capita Gross Domestic Product (GDP) fell from $12,740 to $7,140 in Bahrain; from $22,480 to $16,600 in Kuwait; from $11,450 to $6,725 in Saudi Arabia; and from $27,620 to $14,100 in the UAE.[3] Although true figures are skewed by poor census-taking procedures and the large numbers of expatriates often included in Gulf population figures, these numbers nevertheless reflect the general decline of individual wealth in the Gulf states. As the price of oil fell in the 1980s, Gulf citizens saw their subsidies from governments decline and their high-paying jobs evaporate. In 1994, for example, only a third of the graduates from Saudi universities managed to find jobs in the public sector.[4]

Compounding the resentment created by disappointed expectations is the profligate royal spending found in the Gulf. In Bahrain and

---

[3]Cordesman, 1997a, p. 6, Table Four. These figures are in 1994 dollars, but they are imprecise, because they include expatriate workers in the total population, which skews the true figure for Gulf citizens greatly.

[4]Sick, 1997, p. 17.

Saudi Arabia, residents resent the conspicuous consumption of many royal-family members. Most of Saudi Arabia's perhaps 20,000 princes and princesses receive a stipend from the Saudi state, ranging from thousands to millions of dollars per month.[5] Although these income figures may be exaggerated, Saudis throughout the Kingdom regard them as true. Even more troubling, Saudi royal-family members encroach increasingly on the private sector to maintain their standard of living, taking more government contracts for themselves or demanding a role in other families' businesses while businessmen complain there is less money to go around. In the 1970s, when skyrocketing oil prices seemed to promise enough wealth for everyone, royal interference occurred less often and was more tolerable. Today, there is less money but there are more princes. Bahrain's Al Khalifa, while fewer in number, also maintain an extravagant lifestyle and are perceived to interfere regularly in business for their own enrichment.[6]

The UAE's and Kuwait's huge oil reserves allow them to buy the goodwill of their small populations. Indeed, a repeated theme of all interlocutors is that life in these countries is good. As one government official noted, Kuwaitis and Emirians have "a third half of the pie" to go around. Moreover, by Gulf standards, the ruling families of these states are relatively free of corruption.[7]

A Grim Economic Outlook

As the economies stagnate, resentment over corruption and wealth disparities has become more acute. The Gulf states are hardly poor (and some, such as the UAE and Kuwait, remain extremely wealthy), but the prospects for an ever-rising standard of living are dim. In Saudi Arabia and Bahrain, the number of well-paying, high-status jobs has fallen as the populations have grown. Kuwait and the UAE have large enough reserves to satisfy the wants of their populations for decades to come, but only the UAE (and, in fact, only the Emirate

---

[5]This figure includes all princes and princesses, even those from minor branches of the family, who receive some state money. Simon Henderson provides a similar figure in *After King Fahd*, 1994, p. 7, note 1.

[6]Authors' interviews.

[7]Authors' interviews.

of Dubai) has a significant non-oil trade and banking sector. Although the decline in the price of oil is largely to blame for the economic stagnation, opposition groups rightly note that the regimes squander available resources and fail to develop non-oil sectors.[8]

Some of the dissatisfaction stemming from economic problems is centered on the foreign workforce, which accounts for well over 50 percent of the population in most Gulf states. In general, the number of foreign workers in the Gulf has not diminished significantly, despite rising unemployment among citizens. Employers often prefer to hire foreign workers, whom they pay less, who are better trained, and who generally are willing to work harder. As one Saudi businessmen argued, "Why would I want to hire a Saudi? By law, I must pay them more, I must give them an expensive package of benefits, and if they turn out to be poor workers, I can't fire them."[9] In every Gulf state, the political elite talks about replacing foreign workers with citizens, but little or no progress has been made on this issue over the past decade, and the numbers of foreign workers may even be increasing.[10] Resentment has led to political violence. In Bahrain, for example, many of the arson attacks carried out in the summer of 1997 were on businesses operated by foreign nationals, whom Bahraini groups blame for taking jobs away from citizens.[11]

Bahrain is particularly vulnerable to economic grievances. Once the financial center of the Gulf, Bahrain, because of sporadic violence and government corruption, has lost this role to Dubai, which is seen by many investors as more stable and less corrupt. Bahrain is now in a grim economic situation. Unlike its oil-rich neighbors, Bahrain has almost no appreciable gas and oil reserves. Although Saudi Arabia, Kuwait, and the UAE prop up Bahrain's economy, providing perhaps 55 percent of the government's budget, this support is not enough to prevent steady economic decline. As a result, young Bahrainis are increasingly disenchanted with the regime. Bahrain's Shi'a Muslim community, which is at least 70 percent of the overall population,

---

[8]See, for example, "Your Right to Know: The End of the Deference Era," 1995. Saudi opposition groups blame the Al Saud for the "destruction" of the Saudi economy.

[9]As quoted in Sick, 1997, p. 18.

[10]Sick, 1997, p. 17; Rugh 1997.

[11]Authors' interviews.

bears the heaviest load.[12] Bahraini Shi'a are poor and frozen out of power, while the Al Khalifa and many leading Sunni families maintain a high standard of living and consume conspicuously. Unemployment is over 30 percent among the Shi'a, and possibly significantly higher for young Shi'a males.[13] Unfortunately, the economic outlook for Bahrain is bleak, and resentment is likely to grow. Bahrain has done little to address the core of its problems: corruption, untrained workers, and excessive government interference in the economy. Violence in Bahrain is worsening an already-tenuous economic situation.

Saudi Arabia has far more oil wealth than Bahrain, but the tiny non-oil sector of its economy is doing poorly. Moreover, many Saudis consider jobs involving physical labor unacceptable, believing that it is their right to have undemanding, high-paying, government jobs. The Saudi government, for its part, has shown little inclination to reduce stipends for princes or building of lavish palaces, or to curtail the size of the welfare state.[14]

Saudi Arabia's education system is poor. It cannot educate appropriate substitutes for the foreign-trained technocratic labor force that currently runs the Kingdom. Although the regime's members recognize that they must improve the domestic education system, they are reluctant to challenge the religious establishment, which controls much of the curriculum and fears the import of foreign ideas. Thus, Saudi schools have modern libraries with few books, and computers with no access to the Internet or software. The Al Saud are sending fewer Saudis to study abroad, both because of the expense of a foreign education and because Saudi students often bring back destabi-

---

[12]Historically, in the Gulf states the Sunni sect of Islam has dominated the Shi'a sect. Shortly after the Prophet Mohammed's death, the two sects split over the issue of who would lead the Muslim community. Over time, the Shi'a have developed their own communal identity, which is distinct from that of the mainstream Sunni community. In the Middle East, Shi'ism is dominant only in Iran. Iraq, Bahrain, and Lebanon also have Shi'a majorities; however, in all three countries, the Sunni community dominates political power.

[13]Bahry, 1997; authors' interviews. Unemployment information for the Gulf states is difficult to gather: Individuals have no incentive to register as unemployed, because they receive no financial benefits for doing so. The 30-percent figure for Bahraini Shi'a may be conservative. Fakhro, 1997, p. 177.

[14]Authors' interviews.

lizing political ideas—expectations for liberalism, egalitarianism, or government transparency—that threaten regime legitimacy.

The Saudi government recognizes that change is necessary if its economy is to improve. To date, it has implemented only a few modest reforms. Riyadh has curtailed some defense spending and has extended its payment period on orders for several weapon systems. In 1995, it increased the price of several commodities that were heavily subsidized by the state, such as gasoline and electricity.[15] Nevertheless, defense spending remains high, subsidies to business and on commodities are extensive, and Al Saud princes continue to receive handsome stipends. True economic reform would anger rather than please most Saudis. Although the benefits of privatization and liberalization are long-term, occurring in the form of future jobs and economic growth, the disadvantages occur immediately: Foreign imports swamp domestic producers, streamlined companies lay off workers, and subsidized businesses find they must fend for themselves or perish.[16] Such fears have led the Al Saud to avoid more-sweeping changes.

Brightening this dim outlook will require a change in mind-set.[17] Even as high-paying jobs dry up, the Saudis' contempt for low-status, hand-soiling labor is increasing. Many young Saudis simply wonder why they face worse economic prospects than did their parents in the 1970s, despite being better educated and maintaining Islamic piety.[18]

Corruption and a Lack of Representation

Perhaps the most common grievance Gulf citizens make against their government is the lack of accountability, a lack that promotes abuse of power and rampant corruption. Gulf economies and politics in general are dominated by a few individuals chosen by birth, not

---

[15]Gause, 1997, p. 69; Dunn, 1995.

[16]Gause, 1997, p. 72–73.

[17]However, should the Saudi government impose limited taxes, reduce subsidies and benefits, and otherwise reduce spending it could regain its financial stability fairly easily. Dunn, 1995.

[18]Authors' interviews.

merit. Although the number of ruling-family members in cabinets varies by state, all the important positions—the ones that control spending, internal security, and the military—are dominated by family members or those close to them. Traditional checks on government authority—a free press, an independent judiciary, and a strong civil society (informal associations such as clubs, unions, and other voluntary organizations)—are either lacking or weak in Gulf states. Although several Gulf states have advisory bodies (often called National Assemblies or Consultative Assemblies) appointed by the regime, these bodies, at best, reflect the opinion of elites, and they seldom influence decisionmaking or are able to satisfy popular desires for a true voice in government.[19] The result is widespread corruption and frequent abuses of power, because royal-family members have no constraints on their authority.

The Gulf states, with the partial exception of Kuwait, lack the means to mediate citizen grievances or to ensure accountability. Institutions for organizing political opposition are limited or nonexistent. At times, the governments will anticipate grievances or respond to informal pressures, but such response is generally an ad hoc process. Moreover, the success of a government's anticipation depends on its goodwill; a government may prefer to repress opposition rather than accommodate it, even when the opposition's demands are only for greater government accountability or more public input into decisionmaking.[20]

Corruption and unaccounted-for government spending levels are quite high. Money derived from the sale of oil, noted in balance-of-payment statements, often fails to appear in oil revenues reported in the state budget. In recent years, 18 to 30 percent of the revenue from petroleum exports was not reported in recent budgets in the

---

[19]The distinction between the term "National" and "Consultative" with regard to Assemblies means nothing directly for actual powers but, rather, reflects the Arabic name of these institutions used by the country in question.

[20]Groups that fail to gain popular support for their cause also may become violent. When groups have difficulty attracting new members and have no open means of recruiting, they are more likely to use violence to attract attention (see Della Porta, 1995).

northern Gulf states. This missing money—which totals billions of dollars a year—probably enriches the royal family.[21]

Opposition groups frequently criticize the regime for allowing ruling-family members to mismanage the country. A Bahraini opposition group, for example, accused the ruling family of "squandering the wealth of the nation."[22] Similarly, in a July 14, 1998, communiqué, the Movement for Islamic Reform in Arabia accused the Al Saud members of corruption and blamed them for contributing to the country's economic woes.

Elite corruption is particularly acute and criticized in Bahrain and Saudi Arabia. In Bahrain, the corruption of Prime Minister Khalifa is legendary: He is reputed to demand a cut of as much as 25 percent of any contract of importance, according to several interlocutors. In Saudi Arabia, many royal-family members are well known for their graft. Although these examples are particularly egregious, corruption pervades society, and critics charge that the traditional monarchies are little more than kleptocracies with scepters. All permits, contracts, and other necessary components of business and daily life require bribes to the appropriate officials. Not surprisingly, in Bahrain and Saudi Arabia, both Islamist and secular groups have joined forces to demand greater accountability from the royal family.[23]

The degree of *perceived* corruption probably exceeds its actual levels. Interviews with Gulf citizens and opposition reports suggest phe-

---

[21]Sick, 1997, p. 21.

[22]See "Bahrain: Economy Goes down As Al Khalifa Imports More Foreign Troops," 1997.

[23]One possible instrument of peaceful change in Saudi Arabia is the Consultative Council, the only quasi-representative body in the country—but it is a dim hope. The Council is appointed solely by the King and can legally be dissolved at will by the King. The Council has an advisory capacity, with no authority to oversee government activities or legislate. Optimistic observers might consider that the nucleus of a legislative body exists here; however, in legal terms, the idea of a legislative body contradicts the religious tenets of Wahhabi Islam (and other conservative interpretations), in which man-made legislation is seen in contravention of "God's law," or Shari'a law. Thus, even a proposal to create a legislative body would be dangerous to the regime's legitimacy. There is a small possibility that the regime could eventually move to permit election of members to provincial consultative councils, but more likely it will never permit this dangerous precedent to be established. The King has often publicly stated that democracy is not an "appropriate" form of government for Saudi Arabia.

nomenally high levels of corruption, along with a view that the true reason behind any policy or reform—be it the purchase of an air defense system or an increase in the price of electricity—is the royal-family members' desires to line their pockets. Many Gulf residents appear to believe that eliminating corruption would, by itself, solve their countries' economic problems even though the roots lie much deeper.

Kuwait is a partial exception to the political exclusion and corruption common in the Gulf. Kuwaitis have more political freedom and a more accountable government than do the citizens of other Gulf states. Since its reestablishment after the end of the Gulf War in 1991, Kuwait's National Assembly has served as a safety valve for social pressure. Elected parliamentarians representing all major social groups investigate corruption and oversee a portion of government spending, thus reducing charges of a lack of accountability so common elsewhere in the Gulf. When individuals seek to change society or to oppose a government policy, they now have a legitimate forum in which to express themselves. This option has undercut popular support for both Shi'a and Sunni radicals by providing groups with a voice in and some influence over decisionmaking.[24]

Destabilizing Social Change

In addition to unhappiness stemming from economic and political problems, rapid social change in the Gulf also promotes political unrest. Gulf residents are experiencing a complete transformation of their traditional way of life. As recently as 30 years ago, many citizens of the UAE and Saudi Arabia lived in the desert and had little contact with the outside world. Beginning in the 1970s, these populations were settled, living in modern homes, and depending on the state for their livelihoods. Foreign television shows and movies exposed them to jarring new ideas and ways of life, particularly with regard to gender roles, sexuality, and family relationships.[25] The spread of new

---

[24]Authors' interviews.

[25]Because rapid social change is destabilizing, it is important to note that economic growth also may lead to radicalism among some sectors of the population, even though it may alleviate some disgruntlement over economic stagnation. Currently, many of the unemployed and disenfranchised are sympathetic to calls to use violence

ideas, new forms of communication, urbanization, literacy, and other sources of change disrupted the rhythms of daily life and social hierarchies.[26] Inevitably, some traditional leaders such as tribal shaykhs have lost their influence and almost all individuals face the need to change their ways of life. Not surprisingly, resentment is common, particularly among devout Muslim residents. Even in Kuwait, which faces little Islamist unrest compared with Saudi Arabia or Bahrain, Islamist youths have pressed the government to remove satellite dishes and VCRs in order to fight spiritual pollution.[27]

Issues of social change are constant sources of tension in the Gulf, particularly in Saudi Arabia. Even pious Saudi leaders such as Abdel Aziz and Faysal fought with militants who opposed the introduction of modern devices such as radios and televisions—clashes that often led to violence by either the militants, the government, or both. The smaller and more cosmopolitan Gulf emirates, which are accustomed to limited ethnic and religious heterogeneity, have always been more open to foreign ways than have the more-insular Saudis. Nevertheless, some residents of the small Gulf states also oppose foreign cultural influences. Shi'a demonstrations in 1994–1995 in Bahrain were sparked by Bahrain marathon runners passing through Shi'a villages. Many residents considered their running shorts indecent dress.[28]

Significant disparities in the allocation of wealth and power in society cause another social grievance. When coupled with growing political awareness, such disparities lead individuals to see the government as corrupt or as the servant of a small sector of society. When the pie was expanding, these disparities were more tolerable. In recent years, growing privation has made conspicuous consumption

---

because the system does not provide for them. However, rapid economic growth can shatter traditional institutions and power patterns, leading to the disruption of traditional ways of life, and, in turn, leading some individuals to turn against the regime despite prosperity.

[26]This disruption and dislocation can produce a variety of social effects. Political and social institutions often fail to keep pace with rapid change. Moreover, the change itself can disorient individuals, leading to confusion and anomie.

[27]"Kuwait: Youths Said Seizing Satellite Dishes, VCRs," 1997.

[28]Bahry, 1997.

less acceptable. Such disparities are particularly acute in Bahrain, where the Sunni minority rules over the Shi'a majority population.[29]

In addition to privation, Shi'a citizens in Saudi Arabia and Bahrain endure the burden of widespread discrimination. The Saudi regime has been particularly brutal, restricting Shi'a religious practices and the import of Shi'a religious literature.[30] Shi'a are given instruction in the ultra-conservative Saudi religious doctrine, which preaches that the Shi'a school of Islam is heretical. The Shi'a also face widespread employment discrimination and are excluded from political power.[31] In Bahrain, although discrimination against the Shi'a is less encompassing, the Shi'a are a majority community; thus, their exclusion from political power rankles more. Shi'a in Kuwait and the UAE suffer less discrimination and, as a result, are more loyal to their governments.[32,33,34]

In contrast to many Saudis and Bahrain's Sunni elite, most Kuwaitis are comfortable with other cultural and religious groups and do not feel threatened by sectarian differences. Although 40 percent of Kuwait's population is Shi'a, they are not considered a threat to the regime as are Shi'a in some other Gulf states. Kuwait's Shi'a–Sunni split is far less severe today than Saudi Arabia's or Bahrain's, and the Shi'a in general are loyal to the Al Sabah. In the 1980s, Sunni–Shi'a tensions were quite high as a result of widespread discrimination and the revolutionary ethos being spread by Iran. The Iraqi invasion, and the cooling of the Iranian revolution, brought Kuwaitis together. The Shi'a receive cradle-to-grave benefits; here, no distinction is made

---

[29]Bahry, 1997.

[30]Dunn, 1995; authors' interviews.

[31]al-Khoei, 1996.

[32]Although the above grievances are often attributed to "un-Islamic" sources by religious militants, many of the issues are not "Islamic" or "religious" in the narrow sense of these words. In contrast to much of the unrest in the early 1980s, many of those who oppose the Gulf leadership are inspired by more pragmatic concerns such as corruption and a lack of largesse, rather than concerns related to a strict interpretation of Islam. Indeed, the Saudi regime is regularly criticized by many militants as being un-Islamic, despite the extremely conservative version of Islam it enforces.

[33]Dunn, 1995; authors' interviews.

[34]al-Khoei, 1996.

according to sect. Mainstream Shi'a are represented in parliament, primarily by the Islamic National Alliance (INA). As do other Islamist associations in Kuwait, the INA often works with the government and does not have a revolutionary agenda.

Resentment of the United States

Also provoking resentment is dependence on foreign powers, particularly the United States. Nationalist and religious anger grows when a government openly depends on outsiders for its security. Such dependence can provoke feelings of humiliation among the population and can also discredit the government, leaving it open to criticism that it is unable to provide one of the most basic functions of government: protection of its citizens.

Anti–U.S. opposition groups, even when small in number, often blame domestic problems such as corruption or economic stagnation on the United States. These charges are widely believed, because many Gulf citizens credit the United States with exaggerated influence over domestic politics, as well as over events in the region in general. Popular opposition to U.S. policies, such as the U.S. support for sanctions against Iraq, and resentment of Western-inspired social changes only add to anti–U.S. sentiment.

Because of this perceived link between domestic grievances and U.S. policy, anti-regime agitation or concern about regional politics frequently is expressed through criticism of the United States. Citizens troubled over corruption or the excesses of corrupt royal families might consider U.S. military facilities or personnel an appropriate target for a direct attack, even though the United States, in reality, may have little to do with these problems. Even when violence does not directly target U.S. forces, it can still undermine the U.S. position in the region if it leads to instability among U.S. allies or causes them to limit cooperation with U.S. forces.

The level of opposition to the U.S. presence varies considerably across the region. In Saudi Arabia in particular, both radical and mainstream dissidents have focused much of their protest on the large U.S. presence in the Kingdom.[35] In Bahrain, the presence is often criticized, but not with the frequency and vehemence found in Saudi Arabia. UAE citizens are less concerned with the occasional U.S. presence, and Kuwaitis generally support it strongly.

Saudis are upset about the cost of maintaining the operational U.S. presence and the related arms purchases, arguing that the money could be better spent on services and infrastructure. Much of the business community—many of whom do not strongly oppose the U.S. presence on ideological grounds—criticize U.S. policy in the region because they believe the cost of the U.S. presence has led to a decline in government largesse and is generally bad for business.[36] Opposition figures in exile, such as Usama bin-Ladin, also exploit dependence on foreign powers to highlight the corruption and mismanagement of their own governments. Saudi opposition figures, for example, have noted that the Saudi government has spent hundreds of billions on defense—far more than Iran or Iraq has spent—but is still vulnerable to these countries.[37]

With this resentment comes an exaggerated, but widespread, belief that the United States exercises tremendous influence on—and even control over—internal politics in the Gulf. Thus, many Saudis believe the United States could force the Al Saud to reduce corruption but chooses not to for nefarious reasons, usually related to a purported business conspiracy involving arms sales. Another popular belief is in a grand plan by the United States to keep Saddam Husayn in power as a way of rationalizing the U.S. military presence in the region. Conspiracy theories are advanced to a lesser extent in other Gulf states and range from discussions of British intelligence's orchestration of the death of Princess Diana to a CIA plot to put Khomeini in power to depress oil prices.[38] Such theories abound

---

[35]One of the most famous Saudi radicals, Usama bin-Ladin, declared that, "if someone can kill an American soldier, it is better than wasting his energy on other matters." See "Saudi Arabia: Correspondent Meets with Opposition Leader Bin-Ladin," 1997.

[36]Authors' interviews.

[37]"Escalating the 'Case for Reform,'" 1995.

[38]Authors' interviews.

because the lack of a free press or free government in the Gulf makes it more difficult for individuals to learn what is fact and what is fancy.

Resentment of the United States goes beyond the politics of the day. The legacy of European colonialism—and the subsequent subordination of the Arab and Muslim world to the West—still rankles citizens in the Gulf, many of whom believe the United States, which leads the West today, seeks to keep the Islamic and Arab world weak. Moreover, the United States is held responsible for unwelcome social changes in their country, such as sexual promiscuity or calls for more rights for women.[39]

Many Gulf citizens also disapprove of the U.S. policy toward Iran and Iraq—another reason to be unhappy with the presence of U.S. forces in their country. Except for Kuwaitis, most Gulf citizens feel considerable sympathy for the Iraqi people, if not for their government. They oppose the continuing U.S. sanctions, which are seen as hurting innocent Iraqis while doing little to unseat Saddam. Indeed, many Gulf residents believe that the United States takes pleasure in destroying Iraq and contend that Baghdad no longer poses a security threat to the region. Gulf residents also are skeptical of the U.S. containment policy toward Iran. Many doubt whether Iran truly poses a threat to their security and openly wonder why the United States pursues such a harsh, uncompromising policy toward the Islamic regime in Tehran. U.S. policy toward Iran and Iraq is seldom a top priority for Gulf residents, even for Gulf radicals. Yet the general sentiment that the policy is unfair diminishes support for the United States in general, and for a regional U.S. military presence in particular.[40]

Anti–U.S. radicals often do not distinguish between the U.S. government and local governments. The most extreme simply view area regimes as puppets of the United States. Others blame many of the abusive actions of Gulf governments on U.S. pressure.

Again, Kuwait provides an exception to the common disparagement of the United States. The U.S.–led liberation of Kuwait has engendered tremendous goodwill among most Kuwaitis—goodwill that

---

[39]Authors' interviews.

[40]al-Shayeji, 1997.

lasts to this day, even though the euphoria of the Gulf War period is gone. Kuwait is one of the few countries in the Muslim world where self-described "pro-American Islamists" can be found, although members of the Kuwaiti Hezbollah observe the U.S. presence and may be reporting on it to Iranian intelligence.

Figure 2.4 summarizes the general sources of discontent in the Gulf, by country. Broadly, Kuwait and the UAE face fewer problems than Saudi Arabia and Kuwait.

RANDMR1021-2.4

Grievance	Country			
	Bahrain	Saudi Arabia	Kuwait	The UAE
Expectations/entitlements gap	Moderate Problem	Moderate Problem	Little Problem	Little Problem
No freedom or accountability	Significant Problem	Moderate Problem	Little Problem	Little Problem
Economic problems	Significant Problem	Moderate Problem	Little Problem	Little Problem
Foreign-inspired violence	Moderate Problem	Moderate Problem	Little Problem	Little Problem
Social change	Little Problem	Significant Problem	Little Problem	Little Problem
Dependence on the United States	Moderate Problem	Significant Problem	Little Problem	Little Problem

Little Problem ○   Moderate Problem ⊜   Significant Problem ●

Figure 2.4—Problems Posed by Selected Grievances, by Country

POLITICAL OPPOSITION, BY STATE

This section places the above grievances in context by focusing on active political opposition groups. It describes known major groups in each country and presents an overview of their agendas and concerns. Where appropriate, it also notes foreign involvement in the country's politics.

Saudi Arabia[41]

Opposition in Saudi Arabia is nebulous, but it can be grouped into three streams: disgruntled Shi'a, unorganized Sunni religious militants, and more organized but less influential groups operating from abroad.

Saudi Arabia's Shi'a population is not well organized, but it does have the potential for limited violence.[42] Shi'as rioted several times in 1979 and in the early 1980s, in part to protest discrimination in the Kingdom and in part to show support for the Iranian revolution. The Shi'a are systematically excluded from Saudi Arabia's already-limited political life, controlling no important government positions and having only nominal representation in the Consultative Council. Saudi Shi'a are regularly arrested and harassed by the security services. Moreover, the state does not provide them with their share of government largesse, which represents the key source of wealth in the Kingdom. Perhaps most important, the Shi'a are stigmatized socially. Some regime members and much of Saudi Arabia's Sunni Muslim majority view them as being tantamount to apostates.[43]

Discrimination against the Shi'a has implications for the United States, because the Shi'a are concentrated in oil-rich parts of the Kingdom. Although exact figures are not available, Shi'as probably represent 30 percent of the population of the oil-rich al-Hasa province and as much as 50 to 80 percent of cities such as Hufuf and

---

[41]The intelligence gaps on political opposition groups in Saudi Arabia are huge. Although we have seen little indication that Saudi oppositionists are well organized, this impression may be false. Similarly, the extent to which various grievances are felt is not known.

[42]The Shi'a of Saudi Arabia represent a distinct repressed minority. In Wahhabi eyes, their religion is not even considered as legitimate Islam. Their numbers are much disputed. Most accounts say there are between 200,000 and 400,000; Shi'a claim that half a million of them hide their status and pretend they are Sunni, because of the persecution and disadvantages associated with Shi'ism. In their main homeground, the Shi'a represent 33 percent of the population. The Shi'a themselves say they represent 1 million people, or even 25 percent of the population. This latter figure seems highly suspect and inconsistent (1 million would still be under 10 percent of the overall population). Nevertheless, the claims about large numbers of "hidden Shi'a" are highly plausible. Their numbers are probably between 500,000 and 1 million.

[43]For an overview, see Fandy, 1996.

Qatif. They also represent perhaps one-quarter of the oil industry's manpower.

In the late 1980s, the Shi'a became increasingly pragmatic, focusing on increasing their access to the regime and gaining a voice in the treatment of their community rather than on fomenting revolution.[44] In 1994, Saudi Arabia permitted many Shi'a dissidents to return to the country and allowed some Shi'a to regain jobs in the oil industry, where many had worked in the past but were dismissed from in the 1980s, in exchange for the Shi'a abroad suspending their criticism of the regime's corruption and poor treatment of the Shi'a. The regime also urged Sunni religious leaders to moderate their stance toward the Shi'a.[45] Despite this modest improvement, violence in the Kingdom often brings out the regime's security services; after the Khobar bombing, the government arrested several Saudi Shi'a clergy.

Sunni religious militants, one of the most dangerous groups from a U.S. perspective, oppose social change, ruling-family corruption, and the U.S. presence in the Kingdom. Although information about these groups is difficult to obtain, they probably are not well organized. Any recruitment and organization probably occur through personal or local networks. Some of these groups include individuals who fought in Afghanistan. Such militants carried out the bombing of the Office of the Program Manager for the Saudi Arabian National Guard (OPM/SANG) and may have been involved in the Khobar Towers attack. In addition, they may have a significant following at a local level.[46]

The Islamic opposition in Saudi Arabia can be divided into four groups, although the boundaries blur in practice:

* The followers of religious leaders such as Safar al-Hawali and Salma al-Auda, who gained widespread support after the Gulf War. Many of the Saudis who fought in Afghanistan may support

---

[44]Mottahedeh and Fandy, 1997, p. 308; Fandy, 1996, pp. 5–6.

[45]Fandy, 1996, pp. 6–7.

[46]In 1994, the regime arrested 157 people in Burayda who were protesting the arrest of a religious leader (Dunn, 1995). Opposition groups claim thousands were arrested (authors' interviews).

these religious figures or others with a similar agenda. These are the most important and the least-known groups.

- The Committee for the Defense of Legitimate Rights (CDLR), the Movement for Islamic Reform (MIRA), and other groups that operate from overseas but have little following in the Kingdom itself.

- Saudi Shi'a, who are carefully monitored by the Kingdom's security services and appear to have a nonviolent agenda.

- Radical groups such as al-Qaeda and the Committee for Advice and Reform,[47] which are probably quite small.

Although primarily social, the agenda of these Islamists has a xenophobic, anti-American tinge. The most radical among these groups see any non-Muslim presence in the Kingdom as blasphemy and believe the U.S. presence reflects the Al Saud's corruption. Such groups are particularly incensed by the spread of Western culture, whether in the form of greater freedom for women or the availability of Western media programs in private homes. These Islamists have capitalized on growing resentment of regime corruption and falling living standards, arguing that the U.S. presence is to blame for both.

Even the agendas of more-moderate Saudi groups are quite radical from a U.S. point of view. In a July 1992 "Memorandum of Advice," 107 senior Saudi religious leaders called for severing relations with non-Islamic countries and strictly enforcing Islamic law. They also sought a consultative assembly and a review body for state laws, both of which would be dominated by religious officials.

The only organized Saudi opposition groups are those operating from overseas. The CDLR and MIRA, both based in London, have proven skilled at mobilizing Western government support for their criticisms of the Al Saud on human-rights issues. In addition, these groups have capitalized on information technologies, including the Internet (see, for example, the CDLR's Website at http://www.ummah.org.uk/cdlr/) to spread their message. Both groups draw support mainly from conservative religious figures who are critical of secularist influence in their country and of the Kingdom's alliance

---

[47]Mottahedeh and Fandy, 1997, pp. 307–308.

with the West.[48] However, the groups in exile are removed from the daily events of the Kingdom and have shown little ability to mobilize followers in the Kingdom. Indeed, the CDLR and its offshoots seem to have more influence in London than they do in Riyadh. Also present overseas are mouthpieces for extremely radical Sunni militants.[49]

Bahrain

Bahrain is more vulnerable to unrest than are the other Gulf states. Because the Bahrain Security and Intelligence Service (BSIS) monitors and suppresses all potential opposition, no formally organized and recognized groups operate within Bahrain itself. The ruling family, however, has confronted seething Shi'a masses, who oppose regime corruption, political exclusion, and Bahrain's poor economic state. Available evidence, while limited, suggests that many Shi'a favor substantial political and economic reforms.[50]

The nature of violence in Bahrain is quite different from that in Saudi Arabia: Rather than a few, sophisticated attacks against U.S. forces, Bahrain has suffered many low-level, domestic-oriented incidents, such as arson against foreign-owned shops or stone-throwing. Today, there are no well-organized internal opposition groups in Bahrain. Organized protests are not sustained in the face of government repression, although sporadic unrest remains a problem after a government crackdown.

Bahrain's most-recent political unrest began when elites pressed the Al Khalifa for more accountability. In 1994, about 300 Bahraini elites, both Sunni and Shi'a, signed a petition to the Amir calling for a return of the National Assembly, an elected body with limited decisionmaking authority. (The Amir had dissolved the National

---

[48]Dunn, 1995; authors' interviews.

[49]In London, there are mouthpieces of Usama bin-Ladin, a wealthy Saudi now living in Afghanistan who has supported many terrorist causes. The names of bin-Ladin's groups change regularly, however.

[50]In 1995, 60,000–80,000 Bahrainis gathered in support of a hunger strike by a Shi'a leader—a huge number of demonstrators for Bahrain.

Assembly in 1975.[51]) Shortly after that, some of these elites circulated a second petition, which gathered more than 20,000 signatures. For the second petition, mosques and social meeting places became centers of political activity.[52]

The Bahraini government responded to the petitions by arresting several of the petition organizers and rejecting calls to open up decisionmaking. In the face of this government resistance,[53] violent protests and demonstrations erupted throughout Bahrain. In December 1994, the arrest of Shi'a religious leader Shaykh Ali Salman Ahmad Salman provoked weeks of rioting, and violence flared again in December 1995 and January 1996. During the unrest, crowds of young Shi'a rioted, burning businesses, throwing stones, and otherwise expressing their anger at the government.

Although these Shi'a share the grievances of the petition-signers, their agenda is far broader. Most of the protesters are young, unemployed Shi'a with limited prospects. Not surprisingly, many of the businesses the protesters destroyed were owned by expatriates, who are seen as taking away jobs from Bahrainis. The violent protesters are not organized (in contrast to the petition-signers), and violence is generally limited to arson, breaking windows, and other attacks that do not suggest an organized or highly skilled radical movement is behind them.[54]

So far, these protesters have not focused in any way on the U.S. presence or committed anti–U.S. violence.

---

[51]The petition-signers are heirs to a 30-year trend in traditional opposition. Primarily secular, this trend seeks a return of the National Assembly and involves Sunni and Shi'a elite, who often have Arab nationalist or leftist tendencies. These individuals, often intellectuals, work through petitions or the underground press and are typically nonviolent.

[52]Fakhro, 1997, pp. 181–182.

[53]These protesters are not ideologically driven. Their protests are along communal lines, not religious ones: No religious leader is urging them to go out and protest in the name of Islam; they are acting in the name of the Shi'a community as a people. However, violence does occur regularly after Friday prayers, suggesting the importance of Bahrain's religious infrastructure to many of the protesters.

[54]In the summer of 1997, there were some indications that the opposition was becoming more sophisticated. For example, some of the bombings involved the use of digital timers. And the regime claimed that some individuals were practicing how to use car bombs in their attacks.

In addition to the widespread Shi'a protests, Bahrain does have a limited opposition movement operating from abroad. The Bahrain Freedom Movement (BFM), the largest, best-organized Bahraini opposition group, has moderate Islamic views: It does not demand the imposition of Islamic law or the end of the monarchy; rather, it calls for a return of the National Assembly and a greater distribution of wealth. The BFM's leader is a former member of the religious bloc of the disbanded 1975 parliament.[55]   Like the CDLR and other Saudi groups operating in the West, the BFM has encouraged the West to criticize the Al Khalifa on human-rights grounds.[56] In addition, the BFM has used the World Wide Web (see http://ourworld. compuserve.com/homepages/Bahrain/) to spread its message, and regularly sends out electronic mail to members of the Middle East academic and policy communities. In general, the BFM appears well informed about day-to-day events in Bahrain. Many Bahrainis sympathize with the BFM's goals, and BFM has gained substantial sympathy in the West, particularly in Britain. However, it appears to have little mobilizing power in Bahrain: No protests or demonstrations, even small ones, are in their name. It is rumored that the BFM receives some money from Iran.[57]

Several smaller Shi'a radical groups also operate from abroad, but these probably have little or no influence on Bahraini politics. The Islamic Front for the Liberation of Bahrain (IFLB), the group that almost successfully staged a coup in 1981 with Iran's assistance, has not been active since 1994, and its influence appears to be minimal. The IFLB, which is based in Syria, also has representation in Iran and London. Unlike the BFM, the IFLB calls for the application of Islamic law and the replacement of the Al Khalifa if they will not accept a constitutional monarchy. Shi'a power is their true goal. The IFLB's leading members are in jail, and its presence in Bahrain today probably has little organization.[58]

---

[55]Fakhro, 1997, p. 179.

[56]For an overview of the image the BFM is trying to present to the West, see al-Jamri, 1997.

[57]Also operating from abroad is a group labeled the Human Rights Committee. This group is very small, and its membership is probably limited to a few intellectuals.

[58]Fakhro, 1997, p. 179.

The Institute of Bahrain Studies (IBS), which is based in Iran, disseminates material about repression in Bahrain. Its strength is not known, but it is probably quite small. Iran has helped the IBS with publicity. Available data on the level and type of support provided gave no indications that Iran is turning the IBS into a broader insurgency.

Bahrain also has several once-strong leftist groups that still retain some presence among the elite. The National Liberation Front and the Popular Liberation Front of Bahrain were active in the 1970s in promoting a pan-Arab agenda. These groups helped coordinate a petition to the Amir in 1992 and 1993.[59] They are currently headquartered outside of Bahrain.

Iran tried to create and organize a Bahraini Hezbollah organization, which would spread Iran's Islamic revolution to Bahrain, before and during the 1994–1996 spate of violence, but the BSIS quickly infiltrated and suppressed the organization. Ironically, this group was not linked to any of the violence, although it was active in political agitation. In 1996, Bahrain arrested 44 citizens accused of acting on Iran's behest. Today, the Bahrain Hezbollah probably retains limited organizational capabilities in Bahrain itself, and it almost certainly has some organizational capacity in Iran.[60]

The UAE

The UAE faces little discernible internal opposition. The federation's tremendous wealth allows the regime to buy off any economic problems and satisfy the aspirations of most residents. UAE President Zayid also is widely respected as a proven leader who is both wise and austere relative to the other three north Persian Gulf countries.[61] Although political activity is restricted and there is no formal manner for UAE citizens to influence government, the regime does give the population tremendous economic freedom, minimizing interference in business and allowing merchants to flourish. A consultative

---

[59]Fakhro, 1997, pp. 179–180.

[60]Authors' interviews, conducted in October 1997.

[61]Rugh, 1997. President Zayid lives in small and modest palaces by Gulf standards. He also is known to drive his own car and otherwise have "the common touch."

council at the federal level allows leading UAE citizens a small voice over a few, noncritical government areas. The UAE has no significant radical Islamic movement, no identifiable opposition groups, and little tension between its Sunni majority and Shi'a minority.

Iran is highly active in the UAE, but not in anti-regime activities. Approximately 200,000 Iranians live in the UAE, mostly for business reasons. Iran has access to Dubai as a source of goods embargoed by the West and thus has little interest in stirring up trouble in the federation. Iranian intelligence monitors U.S. personnel in the federation. Despite a great potential for disruption, Iran has not actively meddled in UAE politics. To ensure the federation's relative tranquillity, the security services of the UAE regularly watch the Iranian population.[62]

Kuwait

Kuwait offers an exception to the generalization that the Gulf states lack viable political organizations. Kuwait has large networks of volunteer groups, trade unions, religious organizations, and professional associations: During the Iraqi occupation, these groups played a vital role in maintaining social services and supporting civil disobedience. These same groups are often associated with the political opposition. Kuwait is home to legal political associations and to a comparatively vibrant civil society.[63] Kuwait has three Islamist associations: the Islamic Constitutional Movement (ICM), the Islamic Popular Alliance (IPA), and the Islamic National Alliance. In the 1996 elections, these forces won 15 seats (by some counts).[64] Kuwait's secular leaders come together in the Democratic Forum. The Democratic Forum is a secular-oriented group of former Arab nationalists who seek greater government accountability and an increased popular voice in decisionmaking. Many of its leaders had

---

[62]Authors' interviews.

[63]Article 44 of the Kuwait Constitution allows only the formation of associations, not political parties. After 1991, however, many associations have become, in essence, political parties.

[64]All three groups elected several deputies directly affiliated with them. Six Sunni Islamists who were nominally independent but have ties to both the ICM and the IPA also were elected.

staunchly opposed the United States on Arab nationalist grounds before the Gulf War, but now welcome the U.S. presence. As with the Islamist groups, their primary concerns are social and economic: They seek less corruption, but they oppose the Islamist associations' calls to make society more conservative.

Informally, there are large numbers of gathering places *(diwaniyyas)* where Kuwaitis regularly discuss politics. Kuwait also has legal associations, woman's rights groups, and other organizations that play a role in the political debate, and Kuwaiti labor unions claim thousands of members and are an important base for secular forces in Kuwait.[65]

The ICM and the IPA do not seek radical reform; they focus instead on reducing corruption and instilling conservative Islamic mores. But these Islamic groups are hardly a united bloc. The ICM has ties to the Muslim Brotherhood, a Sunni fundamentalist organization founded in Egypt and existing in different forms in much of the Arab world. The ICM is not revolutionary, but conservative, trying to *preserve* the traditional social order. Its primary agenda is creating a more socially conservative state and trying to reduce government corruption. In the past, the government of Kuwait has worked with members of the ICM to counter leftist groups. The Islamic Popular Alliance, better known as the Salaf (Ancestral) movement, is another Sunni group. More marginal than the ICM, this group calls for a strict interpretation of the Koran in daily life.[66]

The Kuwaiti Hezbollah—a term used indiscriminately to include Shi'a ideologues, those with pro-Iran sympathies, and Islamists who oppose a U.S. military presence in the region—is more a state of mind than an organization, and it poses little threat to the public order. The Kuwaiti Hezbollah's true membership size, while unknown, is probably quite small. In theory, Kuwaiti Hezbollah members seek to establish an Islamic Republic in Kuwait. However, like other Islamist groups in Kuwait, their primary concerns are more social than political. They express anger at the spread of Western values and a concomitant decrease in piety in Kuwait. Neither a well-armed

---

[65]Ghabra, 1991.

[66]Ghabra, 1997.

nor well-trained group, only a minute fraction of these individuals are violent.

A small potential exists for the Kuwaiti Hezbollah to use violence against the regime or against U.S. forces in the region. Some members have access to ordnance left behind by the Iraqis.[67] Prior to the 1992 elections, some Islamists were caught harboring weapons and explosives. Islamists were assumed to be responsible for several explosions that occurred in video stores, which may have been part of a protest against corrupting foreign influences. The Kuwaiti Hezbollah also has ties to groups in Saudi Arabia, Iran, and Lebanon, many of which are quite open.[68]

Kuwaiti politics and political associations also are heavily influenced by tribal politics. Large, wealthy, and prominent tribes such as the Awazem and the Mutairi hold half a dozen seats in the National Assembly. These tribal alliances offset ideological and other bonds.[69]

The government of Kuwait is willing to accommodate Tehran, which may lessen Kuwait's vulnerability to political violence from pro-Iranian Shi'a groups. Kuwait seldom openly opposes Iran, and when it does it ensures that all the Gulf Cooperation Council (GCC) states are behind it. Given the unrelenting hostility of Saddam Husayn's Iraq, the Al Sabah see Iran as a necessary, if hardly ideal, counter to Baghdad's geopolitical influence in the region. Thus, they are willing to negotiate and trade with Iran—and pay Tehran the respect its rulers think it deserves—even though they cooperate with the United States.

Figure 2.5 presents a simple overview of general sources of discontent in the Gulf, by political group. As the figure makes clear, the particular grievances of the groups differ markedly. Some of the groups listed in Figure 2.5 are not formal organizations but, rather, individuals who form small groups or loose alliances, such as the Saudi Sunni radicals.

---

[67]Authors' interviews.

[68]Authors' interviews.

[69]Mottahedeh and Fandy, 1997, p. 305.

RAND*MR1021-2.5*

Opposition Group	Grievance					
	Expectations gap	No freedom or account-ability	Economic problems	Foreign-inspired or foreign-backed violence	Unwanted social change	U.S. dependence
Bahraini petition-signers	⊜	●	⊜	◯	◯	◯
Bahraini Shi'a protesters	⊜	⊜	●	◯	⊜	◯
Bahraini radicals	⊜	●	⊜	⊜	⊜	⊜
Saudi Sunni radicals	⊜	⊜	⊜	⊜	●	●
Saudi Shi'a	◯	⊜	⊜	◯	◯	◯
Saudi exile groups	⊜	●	⊜	◯	⊜	⊜
Kuwaiti Hezbollah	◯	⊜	◯	◯	⊜	◯

Little Problem ◯   Moderate Problem ⊜   Significant Problem ●

Figure 2.5—Level of Problems Posed by Selected Grievances, by Opposition Group

Chapter Three

FROM OPPOSITION TO RADICALISM

Social, economic, and political grievances are common in the Persian Gulf. But grievances alone do not always lead to violence. Despite the atmosphere of discontent discussed in Chapter Two, few individuals in the Gulf actively oppose the ruling families or the U.S. presence, and even fewer use violence to advance their agendas. Understanding what motivates these few radicals requires recognizing factors that politicize individuals to embrace violence,[1] factors that cause politicized individuals to organize to be most effective in a region where few opposition figures are organized, and potential events that could trigger sudden anti-regime violence.

DANGEROUS POLITICIZING EVENTS

Politicization may be an inevitable part of modernity. As literacy spreads, access to different ideas and cultures grows, and traditional patterns of life break down, individuals often engage more with their political systems. Many times, this process is largely or entirely peaceful: Individuals—new actors—become more involved in political systems that gradually embrace them. Governments become

---

[1]Politicization, like terrorism, is a term with many definitions. In general, *politicization* refers to the process whereby apolitical individuals become interested in influencing politics. This process can occur in a benign way, as when passive individuals organize to elect a local school board. It can, however, also be destabilizing, particularly when individuals are politicized in a manner that encourages them to use violence rather than peaceful means to influence politics. For work on politicization, see Gurr, 1993; Goodwin and Skocpol, 1989; Skocpol, 1979; Muller, 1979; Tilly, 1978; and Brush, 1996.

broader-based and take on a greater role in daily life. But politicization can sometimes turn violent. Although a range of factors can lead to violence, four factors are particularly worrisome in the Gulf:

- **Political alienation.** Many Gulf residents believe their political systems are inherently corrupt and unrepresentative. This alienation might lead them to see violence as the only way of influencing Gulf regimes.

- **Defense of traditions.** The rapid pace of social change alienates many Gulf residents, particularly many devout Gulf Muslims. The perceived material and spiritual corruption can lead residents to turn to violence out of frustration or out of anger.

- **Glorification of violence.** The Gulf press and governments have for years praised the deeds of Islamic and Arab fighters in Afghanistan and Lebanon, and against Israel. In addition, many writings of Arab and Muslim intellectuals, particularly among the Islamist community, sanction or even endorse the use of violence to further a political agenda.

- **Interference by Iran and Iraq.** Outside powers can inspire, organize, train, and equip potential radicals, enabling them to use violence effectively. Iran and, to a lesser extent, Iraq have tried these activities in the Gulf in the past and may do so again.

Political Alienation

Violent political action is particularly common when the group or individual in question considers the political system to be completely unrepresentative and believes that it offers no hope for peaceful redress of grievances. Such sentiments of alienation are most common when particular social or communal groups are excluded completely from power. In such circumstances, individuals turn to violence when, with a more representative or inclusive system, they might have used peaceful means.[2]

Political alienation is perhaps the most important politicizing factor for political violence in the Gulf. Because the ruling families domi-

---

[2]Muller and Jukam, 1983.

nate politics, many Gulf citizens perceive their political systems as exclusive. Bahrain opposition groups have called their country a "tribal dictatorship," directly attacking the Al Khalifa family's domination of the country.[3] Even in Kuwait, opposition newspapers have criticized the preferential treatment accorded to the ruling family.

Government unwillingness to broaden decisionmaking increases alienation. After the failure of the 1994 petition drive, for example, many Shi'a in Bahrain concluded that the Al Khalifa would not heed peaceful voices. And, indeed, many among the Al Khalifa are opposed to granting any concessions to opposition figures.[4] Saudi Shi'a also suffer extreme political exclusion. Not surprisingly, interviews indicated that many Gulf citizens believe that peaceful political activity will not influence their country's leadership.

Because they are unable to influence decisionmaking, even peaceful citizens may become frustrated and come to believe that only violence will move the regime to action. This opposition could have dangerous repercussions in view of the moderate political groups throughout the world that have frequently become violent after years of repeated failures in proposing compromise.[5]

Government crackdowns on peaceful opposition groups also increase alienation. The harassment and violence that often accompany a crackdown can lead to greater levels of resentment. On a practical level, opposition members who are forced underground after a crackdown may turn to radicals for help in sustaining themselves and their families. Perhaps most important, a government crackdown demonstrates the intolerance of a regime to its citizens, representing symbolic exclusion that, in and of itself, is a potential grievance.

Getting a voice in government can lessen this problem. In Kuwait, although the National Assembly and the government are often at odds, the opposition's voice—and its modest ability to affect policy—

---

[3]"Bahrain: Alleged Conspiracy Used As a Cover for Consolidating Tribal Dictatorship," 1996.

[4]Authors' interviews. In Kuwait, this sentiment is countered somewhat by the limited redress offered by the parliamentary system.

[5]Della Porta, 1995.

have increased the overall level of support for the Al Sabah.[6] Because Kuwaitis can organize peacefully and effect political change, there is less reason to embrace violence. Bahrain, on the other hand, cracked down on dissent, and a petition drive for greater representation led to widespread street rioting.

An issue of concern neglected by Gulf regimes is that political but nonviolent individuals might turn to violence as a result of a regime crackdown on mainstream politics. Except in Kuwait, the Gulf ruling families are quick to suppress most forms of political expression. Particularly in Bahrain and Saudi Arabia, even oppositionists seeking fairly benign reforms, such as greater respect for civil liberties and more government accountability, are subject to potentially brutal regime countermeasures. As a result, reformers as well as radicals have at times had to go underground or to flee abroad—actions that make them more susceptible to becoming a clandestine movement that would use violence in lieu of peaceful protest.

Political alienation is particularly dangerous when it happens to elites. As noted in Chapter Two, the ruling families dominate politics in the Gulf and, to an increasing extent, control Gulf economies. Newly educated technocrats often are angered by the rampant corruption and inefficiency in the Gulf and question why the country is controlled by poorly educated family members who have few qualifications for office. Similarly, religious scholars resent the state-sponsored, and often intellectually inferior, religious leaders who dominate their countries' religious establishments.[7]

Defense of a Traditional Way of Life

When individuals perceive that the demise of the status quo is imminent, they are more likely to undertake risky actions to maintain or improve it. Doing nothing, in such situations, is perceived as tantamount to accepting the loss of the old status quo and its advantages. Social change in traditional societies often produces this concern. Rapid modernization frequently disempowers traditional elites, such as those affiliated with leading families or religious leaders. Such

---

[6]Authors' interviews.

[7]Gause, 1994; and Crystal, 1995.

individuals see the new class of technocrats, who received their positions through education rather than birth, and the general rise in state power as a threat to their position and privileges.

Many establishment voices decry the transformation of their societies. Likewise, many Gulf radicals believe that they are defending their way of life against the onslaught of Westernization and secularism. As noted in Chapter Two, the Gulf has changed dramatically in the past four or five decades, going from a poor, nomadic society to a wealthy, settled one. Similarly, technology and international trade have connected the Gulf states to the world at large. In addition, the flow of oil wealth into the state's coffers has enabled the ruling families to shunt aside members of traditional merchant and tribal elite families. These families resent the "upstart" royals who were their equals, or at times even their inferiors, several generations ago.

The Glorification of Violence

Interviews with arrested terrorists in a previous RAND study suggest that many individual terrorists are motivated by the deeds and practices of violent role models or the teachings of intellectuals who condone or support the use of violence. Although every group and society can find some violent forefather to claim as their own, some societies and cultures are more willing than others to glorify violent individuals.[8]

The glorification of individuals who use violence is common in the Gulf. For years, stories of brave Palestinian *fedayeen*, willing to risk their lives to recover Arab and Muslim lands from the Zionist invaders, nurtured much of the Arab and Muslim world. The successes of the Lebanese Hezbollah and Bosnian Muslim fighters also became the stuff of legend, and many fighters were lionized in the Arab media. Such publicity inspired activists to become radicals by proving that violence succeeds. It also suggested a potential source of social esteem, because the fighters in these causes (some of whom were Gulf citizens) were lauded.

---

[8]Kellen, 1979.

The anti-Soviet fighting in Afghanistan also mobilized many Gulf citizens. With the active encouragement of their governments and religious leaders, many Gulf youths joined the anti-Soviet Islamic fighters in Afghanistan. There, they learned both how to fight and how to organize a political movement. They also gained a network of well-armed, transnational contacts who share a fervent Islamist political mind-set.

Violent activists also respond to the intellectual environment of their communities. Because many radicals are well educated, they will often turn to works of philosophers, political theorists, and theologians in seeking guidance for their actions. When these works explicitly call for violence—or, at the very least, do not condemn it— individuals are more likely to use violence in their actions.[9]

The intellectual environment of the Gulf is favorable to radical causes. The dominant *zeitgeist* of the 1960s and 1970s was Arab nationalism, whose proponents called for the Arab masses to turn against their regimes and the Western powers. Seizing power through violence, whether through a military coup or a bloody revolution, was the proffered model. While still strong in some circles, Arab nationalism—the doctrine that the Arabs share a common political identity regardless of state boundaries—has faded since the 1970s, and radical Islamic sentiments have grown stronger.

But radical Islam also endorses violence in politics. Although the doctrines of radicals vary, some influential theologians have in essence declared certain regimes "heretical," implying that the faithful should overthrow them by any means possible. Theologians often play a direct role in the formation and direction of radical groups.[10]

Interference from Outside Powers

As will be discussed further in Chapter Four, Iran, Iraq, and nonstate groups such as Hezbollah sometimes try to influence Gulf domestic politics. Iran in particular has encouraged radicalism, most recently

---

[9]Kellen, 1979.

[10]Ajami, 1991; Dabashi, 1993; Green, 1982; and Landau, 1994.

in Bahrain, and has tried to organize opposition groups capable of using violence. Such outside aid makes individuals more likely to use violence, because their vital organizational assets—funding and a haven overseas—depend, at least in part, on following the wishes of an outside power. Such aid helps them triumph over less-radical rivals and survive the repression of the state. Ironically, the Gulf regimes' very skills at suppressing dissent increase the importance of outside supporters. Because it is difficult for a large, well-organized group in the Gulf to survive for long, a haven overseas and outside assistance become more important. Indeed, many groups, particularly Shi'a opposition leaders, have historically begun organizing their communities overseas for this very reason. In recent years, Sunni radicals have also begun organizing themselves overseas after being expelled from their Gulf homes.

As Figure 3.1 reveals, violent politicization is much more of a serious problem in Saudi Arabia and especially in Bahrain than elsewhere in the Gulf. In these states, the combination of politicizing factors keeps the potential for anti-regime opposition strong.

RAND*MR1021-3.1*

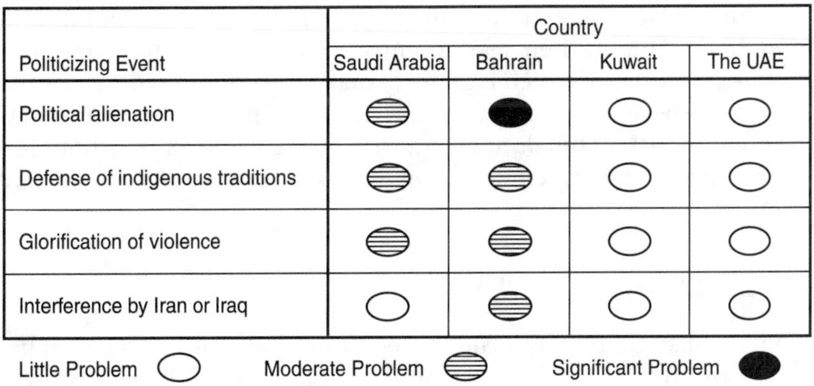

Politicizing Event	Country			
	Saudi Arabia	Bahrain	Kuwait	The UAE
Political alienation	⊜	●	○	○
Defense of indigenous traditions	⊜	⊜	○	○
Glorification of violence	⊜	⊜	○	○
Interference by Iran or Iraq	○	⊜	○	○

Little Problem ○   Moderate Problem ⊜   Significant Problem ●

Figure 3.1—Problems Posed by Politicizing Events, by Country

Implications for Countering Violence

Drawing on deep emotional currents or extra-regional events that governments cannot control, several of the politicizing sources are extremely difficult for Gulf states to counter. Most obviously, Gulf governments find it hard to delegitimize the use of violence for political purposes. Gulf ruling families spent years trying to weaken Arab nationalism, but only Israel's repeated defeats of Arab nationalism's champions and the economic stagnation it produced discredited this ideology among much of the intelligentsia. In general, the promotion of more-peaceful role models is probably beyond the scope of a lone regime. Halting anger of the threat to traditions also is difficult. Although Gulf governments, particularly the Saudi regime, are trying to resist outside influences and to preserve a traditional way of life, new ideas still penetrate the Gulf region, and social change continues at a rapid rate.

Regime impact will probably be strongest in determining whether individuals turn to violence because a regime cracks down on more-moderate opposition groups in general. By suppressing all opposition, Gulf regimes may be forcing reformers to become revolutionaries. For opposition movements to survive in Bahrain and Saudi Arabia, their leaders must go underground or organize overseas—either of which is likely to make them more radical. As the Kuwaiti experience suggests, some degree of political accommodation could offset this danger, but the ruling families in other Gulf states have so far proven unwilling to make more than token political reforms.

ORGANIZING OPPOSITION

Political instability in the Gulf differs from that in other regions, because the region's citizens seldom organize themselves in formal, well-institutionalized movements. Individuals who cannot organize find it harder to press their governments or otherwise affect politics. Because of constraints on political organization, religious organizations often are disproportionately important. However, organization is generally limited, which reduces the effectiveness of political opposition and also makes political violence less effective.

Impact of Opposition Organization

The ability to organize shapes both the propensity for violence and the type of violence that *is* carried out. When governments permit groups to organize, it makes those groups more potent political actors, but, by giving individuals another, peaceful means of affecting decisionmaking, it also reduces the likelihood of violence. In the Gulf, most organization, both social and political, occurs informally, through religious and personal associations. Because formal organization is limited, grievances are often expressed spontaneously, through demonstrations and violence.

Political organization is a key variable to monitor when assessing the potential for political violence and the effectiveness of counter-political violence, measures discussed in Chapter Five. Terrorist groups are more effective when they are organized, but they are also easier to infiltrate and destroy. However, when there is no formal structure or when cells make independent decisions, it is difficult for intelligence agents to penetrate and control those organizations. Knowing the members and actions of one local group often offers little insight into the behavior of groups with a similar ideological inclination but no formal relationship.[11] The result is a less effective group, since it is harder for it to coordinate strikes, but, so too is it harder for it to be targeted.

Organization affects the range of threats that terrorist groups pose. Almost any armed individual can threaten the lives of unprepared victims, whether they be local nationals or U.S. soldiers. A large number of unorganized individuals also can create a climate of unrest in a country. However, a small but organized group that can coordinate attacks and strike the most-important targets can also effectively create a climate of unrest. Small and organized groups also can attack well-protected bases or buildings. Most difficult of all is coordinating with opposition armies, which requires a high degree of training and sophistication. Few militaries in history have conducted coordinated operations with terrorist groups, preferring to use highly trained special operations forces to gather intelligence,

---

[11]Wardlaw, 1990.

disrupt adversary communications and supply, and otherwise operate behind enemy lines.

Limited Autonomous Organization

Gulf governments, with the exception of Kuwait and possibly the
UAE, seek to keep any autonomous political organization limited. If
a group openly recruited for membership or sought financing, the
police would quickly arrest its members. When people do organize,
it is usually in small, local gatherings. Even mosques are monitored
by security services; thus, most political activity occurs in private
homes and in private gatherings.[12]

Bahrain's 1994–1996 spate of oppositional political activity illustrates
how poor organization generates resentment while hindering effective political opposition. Organization in Bahrain, when it occurred
at all, was conducted through informal gatherings: Any regular
meetings were quickly infiltrated and suppressed by the BSIS.[13] The
lack of any mechanism such as an elected assembly to influence
decisionmaking or express grievances increased popular frustration
with the regime and led elites to demand the restoration of the
National Assembly. Because there was no institutionalized, formal
way for elites to press the government on this point, they issued
petitions. The failure of the petitions led to demonstrations and
riots, but not to a sustained campaign of resistance. At each stage, a
lack of organization shaped the form and strength of the protest.

In addition to lacking formal institutions for political organization,
most Gulf states lack a strong civil society from which political opposition might develop. In Saudi Arabia and Bahrain, the government ruthlessly suppresses informal organization, fearing that it
might lead to political activity. Although the UAE does not systematically suppress all civil society, the federation has few informal discussion groups, unions, or other vestiges of civil society, and many

---

[12]Authors' interviews. Shi'a groups in particular have little ability to organize in Saudi
Arabia. However, in 1993, the government made peace with some radical Shi'a, letting
leaders return to the Kingdom in exchange for amnesty. At the same time, the
government made token gestures such as increasing support for Shi'a meeting halls
and schools.

[13]Authors' interviews.

theoretically independent organizations—such as Chambers of Commerce—have their leaders appointed by the government. Economic dependence on the regime limits an important form of organization—the private business relationships and associations that often form the nucleus of future political networks. But the state-dominated nature of Gulf economies makes such networks less important, forcing individuals to work through the state rather than independently.

Religious Organization

In general, religious groups have shown more-impressive organizational skills than have other oppositionists. Religious groups have an advantage: They can work through existing channels that receive the ostensible support, or at least the toleration, of the regime. Religious organizers often exploit existing religious and social meeting halls, which are necessary for day-to-day religious activities.[14] Religious leaders, particularly Shi'a religious leaders, have informal ties through common educational experiences in religious academies in Iraq or Iran. In addition, religious groups often are stronger and better organized because many Islamists are highly committed and are willing to work for ideology as well as for material benefits.

Shi'a groups often find it easier than do Sunni groups to organize autonomously. Ironically, because all the Gulf regimes favor Sunni Islam, Sunni religious leaders receive government salaries and more support in general—support that makes them more subservient to the state. Their Shi'a counterparts, in contrast, rely on the Shi'a community for funds, which gives them considerable independence. This autonomy is common to Shi'a groups throughout the Muslim world, not just to those in the Gulf.

Shi'a Islam also is more hierarchical than is Sunni Islam, making co-ordinated action easier. The Twelver school of Shi'a Islam—the dominant school in Iran and among Shi'a in Saudi Arabia—separates the religious leadership from the Shi'a community as a whole, a sep-

---

[14]As Ayatollah Hussein Ali Montazeri, a leading Iranian religious leader involved with Shi'a radicals abroad, noted: Mosques "should not only be places of prayer but, as in the Prophet Mohammed's time, should be centers of political, cultural, and military activities." As quoted in Hoffman, 1990, p. 5.

aration that, in effect, creates a class of religious leaders. Furthermore, the Shi'a scholarly community recognizes loose but widely accepted distinctions based on the degree of learning. Such distinctions often result in an informal, but quite effective, religious hierarchy. When a scholar is highly charismatic and highly political—such as the late Ayatollah Khomeini—his followers often form the core of a political movement. As with Khomeini, widespread emulation by followers and disciples can give the scholar tremendous influence, even outside his home country. This religious emphasis on hierarchy and transnational ties—Shi'a from all countries have studied in the Iraqi cities of Najaf and Karbala and the Iranian city of Qom—means that coordinated action among Shi'a worldwide is easier. However, many Bahraini Shi'a are of the Akhbari school within Twelver Shi'ism, a tradition that de-emphasizes the rationalist orientation of the mainstream Usuli branch of Twelver Shi'ism and rejects the right of religious leaders to flexibly interpret Islam.[15] This tradition, in effect, reduces the ability of religious leaders to sway the masses of Bahraini Shi'a to new causes.[16]

Shi'a organization frequently takes place overseas. Iran and Hezbollah often capitalize on their religious and cultural ties to Gulf residents and organize those residents when they leave their home countries. Many leaders of Bahrain's Shi'a opposition—including both moderate reformers and more-radical Islamists—are led by younger Shi'a religious figures who studied together in Qom. Although not always acting directly on Iran's behalf, these scholars observed a model of religiously organized and inspired activism in Iran that they brought back to Bahrain. There, both through deliberate inculcation and observation, they adapted much of the Iranian revolution's religio-political activism. Shi'a from the Gulf states also regularly travel to Lebanon, where they meet other religious Shi'a.[17]

---

[15]The Twelver school of Shi'a Islam endorses *ijtihad*, the use of reason and the principles of jurisprudence to arrive at judgments. *Ijtihad* can give senior religious leaders tremendous influence, because they can make judgments about a wide range of daily activities, including political events. However, the concurrent belief in *ihtiyat* (prudence and caution) limits the more excessive use of *ijtihad* in practice (Momen, 1985).

[16]Momen, 1985.

[17]Bahry, 1997; authors' interviews.

Given problems with organizing either formally or informally, it is not surprising that spontaneous political action is more common in the Gulf, and in the Middle East in general, than in the West. Understanding when such violence will occur is less a question of knowing the plans of specific groups than of anticipating potential events that will set unorganized violence in motion. The unorganized nature of political violence in the Gulf may cause such violence to be part of a new form of unrest: radicals lacking well-defined objectives and with motives that are sometimes not easily comprehended.[18] Thus, by exploring potential triggers for violent action, the next section focuses on when these individuals might decide to act.

POTENTIAL TRIGGERING EVENTS

Political change in the Middle East is notoriously difficult to anticipate. Major events—the fall of the Shah, the Israel-Egyptian peace, and the Iraqi invasion of Kuwait—occurred fairly quickly, surprising most experts. Nevertheless, several events that could lead to instability loom on the horizon. Changes in regime leadership and destabilizing reforms might result in widespread unrest. As with other Middle Eastern countries, countries in the Gulf also are vulnerable to instability resulting from events outside the region. As with domestic complaints, external grievances can lead to anti-regime and anti–U.S. political violence. Although the potential range of external stimuli is vast, oil politics, the Arab-Israeli conflict, bloodletting in Bosnia, the struggle in Afghanistan, and exposure to foreign cultural influences all have caused resentment among Gulf residents in the past and may do so again, as have political reforms in neighboring Gulf states—changes that have often had a domino effect, leading citizens of one country to demand the freedoms enjoyed by others.

Succession Crises

Several Gulf leaders are widely popular, and their deaths could increase dissatisfaction with their regimes. Moreover, if power rapidly

___

[18]Hoffman, 1998, p. M1.

shifted, it could lead to infighting in the royal families and anger among those left out of power. Splits in the royal families could hearten opposition figures, particularly if the governments appeared weak.[19]

Fortunately, the succession in the Gulf states looks promising. In Bahrain, the long-ruling Amir Isa died in March 1999, but his son and successor Hamad, while less popular, may be more willing to conciliate the opposition and faced no unrest when he assumed the crown. In Saudi Arabia, Crown Prince Abdallah is widely seen as a pious and honest man—far more so than King Fahd. Moreover, he is the long-time head of the Saudi National Guard and has kept up family ties with the tribes from the Najd region, whose members are often active supporters of radical Islamist groups.[20] Thus, he may be more esteemed by the regime's harshest critics. Kuwait's Crown Prince, Saad Abdallah, is popular and may be a more effective ruler than the rather aloof Amir Jaber. Any successor in the UAE will find it hard to fill the shoes of the widely respected Shaykh Zayid, but his son and heir, Khalifa, has gradually developed a good reputation.[21] Gulf leaders in general also recognize that dissension within the ruling families will only weaken the regime as a whole, a factor that favors stability.[22] In general, these potential heirs favor continued good relations with the United States and approve of the U.S. military presence in the region.

A troubled succession might lead to a long period of regime weakness. It may take years for even strong leaders to consolidate power and control decisionmaking. Thus, consensus will be required for all but the most uncontroversial decisions, which will delay or prevent reforms and other politically unpopular measures as the rulers who lack a solid base within their own governments strive to retain popular goodwill. Such weakness and hesitation could become especially problematic during a foreign crisis or internal instability if the leadership hesitates, or refuses, to invite U.S. forces in to defend their countries.

---

[19]Authors' interviews.

[20]Dunn, 1995

[21]Rugh, 1997.

[22]Authors' interviews.

Destabilizing Economic and Political Reforms

In addition to a leadership change, several reforms that would strengthen the Gulf states militarily and economically in the long term could lead to political unrest in the short term. Imposing taxes, for example, would violate the implicit social contract put forward by the ruling families: The ruling regimes will provide cradle-to-grave coverage while asking nothing in return, except for political loyalty and passivity. With taxes might come demands for a greater voice in decisionmaking.

The creation of mass armies might have a similar effect. Once citizens come to play a role in the state with their blood and treasure, they are less likely to stay politically quiescent.[23] Thus, the Gulf states will rely on the U.S. military for the foreseeable future.

Attempts to open the regimes politically also might have a short-term destabilizing effect. In many countries, democratization has been accompanied by political violence. Individuals in the country respond to their increased freedom by organizing against the government, using violence against rivals, or otherwise perverting the ideas of free assembly and free discourse. In the Gulf, political opening might lead to greater organization by Islamists or other groups opposed to the ruling families.

Instability Spreading from Abroad

Events outside the region also could lead to greater instability in the Gulf. The further collapse of the Arab-Israeli peace process could dishearten many residents and turn them against the United States. A victory of radical Islam elsewhere in the Middle East, such as in Algeria, could increase the prestige of Islamists. The rise of a radical religious leader also could radicalize many Gulf Muslims.

Steps toward democracy abroad, particularly elsewhere in the Gulf region, might inspire unrest in states with less-representative governments. Perceptions of political exclusion are often relative:

---

[23]As Gause (1997, p. 66) argues, Iran can gain manpower through ideological fervor; Israel, through democratic ties; and Iraq, through compulsion. The Gulf states have none of these options. Kuwait has obligatory service, however.

Individuals weigh their freedom by comparing it to that of their neighbors. The restoration of Kuwait's parliament after the Gulf War and the 1992 establishment of a consultative council in Oman inspired many Bahrainis to demand their own National Assembly.[24] Similar advances made in another Gulf state may trigger calls for reform elsewhere in the region.

The spread of democracy since the fall of the Berlin Wall also has affected the Gulf. In Asia, Eastern Europe, and other former bastions of authoritarianism, dictators have been replaced by popular regimes. Yet nonliberal regimes remain the norm in the Middle East. The Gulf in particular—perhaps the only region of the world in which monarchy is the established form of government—increasingly appears an anachronism.

Setbacks in the Arab-Israeli peace process or transregional issues involving Muslims have led to tension in the Gulf that could spill over into anti–U.S. violence.[25] As the Arab-Israeli peace process has stalled, so too has the goodwill built up by the United States' early recognition of the importance of the Palestinian issue. The United States is seen as Israel's protector and as being able to control Israel's actions. Even in pro–U.S. Kuwait, students have demonstrated against U.S. support for the Israeli government.

Israel is hardly the only cause that can inflame popular sentiment against the West. The killings of Bosnian Muslims appalled Gulf residents, and aid to Bosnia became the favorite charity for many religious organizations. Some Gulf residents compared the situation in the former Yugoslavia unfavorably with the U.S. response to the Iraqi invasion of Kuwait, noting that the United States would not intervene when Christians killed Muslims.[26]

Such pan-Islamic issues resonate in the Gulf, although more so with some issues than with others and according to the fervor of individual citizens. Islamic identity is often stronger than national identity in the Gulf. Thus, when Muslims abroad fight or die, it often strikes a responsive chord in the Gulf. This sympathy does not compel all

---

[24]Bahry, 1997.

[25]Khalilzad et al., 1996.

[26]Authors' interviews.

Gulf citizens to go and fight abroad, but it does lead to widespread moral and financial support for Islamic causes. A few more militant believers may even take up arms themselves.

Chapter Four

FOREIGN INTERFERENCE

The question of Gulf stability and political violence is often asked in the context of Iranian or Iraqi machinations in the region. How does foreign support—also known as *state support*—cause or exacerbate political violence in the Gulf? Foreign support can both increase the overall atmosphere of discontent and politicize individuals against their own government. This chapter examines support for political violence by Iran, Iraq, and various radical groups and notes its potential impact.

Instability in the Gulf is especially dangerous because aggressive foreign powers—particularly Iran and Iraq—might capitalize on it, backing radical groups to serve their own ends. Even when terrorist groups have no relationship with foreign powers, Iran or Iraq might use existing violence as a pretext for intervening or could simply strike when the Gulf states are weakened by internal violence. Outside powers can increase the overall level of disgruntlement in a country. Iran and Iraq both have at times harangued the traditional Gulf monarchies, denouncing their leaders and calling on their citizens to overthrow them. In addition, outside powers can highlight injustices, both real and perceived, making citizens more resentful of their government. The Iranian revolution brought into relief the political alienation felt by the Gulf Shi'a.

Outside powers can also politicize individuals through aid and by example. Foreign governments can issue a call to arms, helping to mobilize and organize a population that was previously passive. Iran, for example, broadcast calls for Gulf Shi'a to rise up after the Iranian revolution; Egypt's Nasser acted similarly when he called for

Arabs to unite behind his banner and throw off reactionary regimes. Individual religious leaders outside the Gulf often inspire individuals in the Gulf, shaping their demands of government and their broader political agenda. Foreign regimes also provide a model for political action, demonstrating that a religious or nationalist movement can realistically aspire to power. Iran's very example inspired many Muslims, both Sunni and Shi'a, to strive for Islamic government. In addition, Iran helped train radicals, using its political model to organize them.

State support can greatly increase a radical group's capabilities and its range of activities. States can provide money, weapons, and training to terrorist groups. Their diplomatic pouches can smuggle weapons and explosives past government security services, and their intelligence officers can provide operational information to make terrorist strikes more effective. After an attack, a foreign government can provide a safe haven for terrorists.[1]

The role of foreign support for Gulf radicals has changed since the early 1980s. In the past, Gulf radicals often acted as proxies for foreign governments. Shi'a groups backed by Iran committed terrorist attacks against U.S. and Gulf targets—such as trying to assassinate the Kuwaiti Amir and blowing up the U.S. Embassy in Kuwait—in part to further Iran's agenda. Today, however, Gulf radicals themselves regularly seek external support to improve their capabilities for expressing their own grievances. Through ties of veterans of the struggle in Afghanistan, religious organizations, and other cross-border networks, Gulf radicals are able to obtain the support of movements and governments abroad. Direct foreign support remains a grave threat: Iran in 1994–1996 tried to stir up violence in Bahrain, and Iraq would no doubt use political violence were it allowed to operate freely (in fact, in 1993 Iraqis attempted to assassinate former President George Bush when he visited Kuwait). Nevertheless, direct support for violence appears to be declining. Gulf residents are increasingly seeking outside support on their own. Whether this is a temporary or a permanent change is not clear, but it does suggest that more attention should be directed at foreign

---

[1]Tucker, 1997.

governments' abetting, as opposed to directing, political violence in the Gulf.

Foreign support also increases the likelihood of political violence being directed against U.S. forces and facilities, because both Iran and Iraq vehemently oppose the U.S. military presence in the Gulf. Foreign support for terrorists can make them far more effective and deadly. In the Gulf, Iran, Iraq, and transnational Islamic movements all might aid local activists against their own regimes and against the United States.

In the past, Iran actively supported political violence as part of its foreign policy, frequently trying to create local proxies to carry out its wishes and to spread its revolutionary credo. In addition, the Islamic Republic has used political violence to assassinate regime opponents, harass rival governments, and demonstrate its commitment to the worldwide Islamic cause. Iran has not directly supported violence in the Gulf since backing Bahraini radicals during the 1994–1996 unrest (and the verdict is still out on who was responsible for the 1996 Khobar Towers attack); however, it *has* tried to create local proxies in the Gulf that are armed and capable of using violence at Tehran's behest.

Iran is not the only foreign actor that might support violence in the Gulf. Iraq regularly employed political violence in the past; given Saddam Husayn's ongoing hostility toward the United States and its Gulf allies, it is likely to try to use political violence to advance its agenda in the future. The Lebanese Hezbollah also has worked with Gulf citizens, inculcating them with radical Islamist teachings and perhaps training and arming them as well.[2]

IRAN'S USE OF POLITICAL VIOLENCE AS A FOREIGN-POLICY INSTRUMENT

Iran has not hesitated to support violent radical groups throughout the world, and it has repeatedly supported militants in the Gulf. Iran

---

[2]The Muslim Brotherhood is active throughout the Middle East, and numbers many teachers among its adherents. In Egypt, the Brotherhood itself has supported peaceful politics in recent years. However, several Brotherhood offshoots, such as the Brotherhood in Syria and HAMAS in the West Bank, are extremely violent.

uses political violence for four main purposes: (1) to build up a local group to act as its proxy; (2) to assassinate regime opponents; (3) to press rival governments; and (4) to enhance its claim to be the vanguard of the Islamic revolution.[3]

In general, Iran prefers to work through proxies to maintain deniability and thus avoid the opprobrium associated with being a known sponsor of terrorism. The Lebanese Hezbollah in particular has carried out many terrorist attacks on Iran's behalf.[4] For example, Iran enlisted elements of Iraq's Al Da'wa movement and the Lebanese Hezbollah to attack U.S. and French targets in Kuwait in 1983. Iran also sent Hezbollah fighters, along with Islamic Revolutionary Guards Corps (IRGC) members, to work with Bosnian Muslim fighters. Tehran, meanwhile, has tried to avoid creating direct evidence that it controls the Lebanese Hezbollah, despite the fact that Iranian government officials formally sat on Hezbollah's directing body for many years.[5] Similarly, when trying to create a Bahraini Hezbollah, Iran stressed the importance of keeping its sponsorship hidden.

A number of organizations and bureaucracies in Iran are active in supporting political violence abroad. Most important are the IRGC and Iranian intelligence, both of which have training programs for activists in Iran and abroad. Quasi-governmental bodies in Iran, such as the revolutionary foundations that control vast amounts of funds, also have their own agendas related to political violence. For example, one foundation, not the Iranian government, has placed a bounty on author Salman Rushdie. Individual religious leaders also have networks of students and followers abroad, and often pursue their own agendas. Although the Iranian government controls—or can control—many of these activities, it is often not clear who in Iran is behind a particular attack.

---

[3]For an overview of Iranian support for terrorism in the 1980s, see Hoffman, 1990.

[4]O'Ballance, 1997.

[5]Jaber, 1997; Ranstorp, 1997.

Proxies: Fostering Insurgencies

Iran's preferred *modus operandi* is to develop an insurgency or popular movement that, over time, will dominate the country. Tehran's ideal is to create a proxy dedicated to the idea of the Islamic revolution; that proxy will eventually gain control over the country in question. However, as the fervor of the Islamic revolution has worn off, the Islamic Republic's leaders also have recognized the value of using local parties to press the regime to support Iran's policies in general. Iran used this method most successfully in Lebanon, where the IRGC built up Hezbollah and made it the leading Shi'a organization in the country.[6]

Although Tehran initially encouraged Hezbollah to assume control over Lebanon and turn the country into an Islamic republic modeled after Iran, today Hezbollah acts in part as does a conventional political movement, seeking greater influence rather than absolute dominance. Iran has also sent hundreds of IRGC Al-Qods Forces, Iran's special forces responsible for training insurgents abroad, to Bosnia, along with members of the Lebanese Hezbollah, to build up support among the Muslim community there.[7]   Iran also has tried similar tactics in Iraq, working with the Al Da'wa Party and the Supreme Assembly for the Islamic Revolution in Iraq to overthrow the Baath regime.[8]

These IRGC-backed movements are more guerrilla groups than terrorist organizations, but the line between the two is frequently blurred in practice. Iran attempted to build up such an organization in Bahrain in 1993–1994, but the BSIS quashed it before it gained widespread support. Iran sought to develop a Bahraini Hezbollah movement that would recruit members and become a viable political organization; its long-term goal was to establish a pro-Iranian Islamic republic. To this end, the Revolutionary Guard's Al-Qods Force trained several Bahrainis in Iran as a local leadership cadre and provided the group with limited financial support. Bahraini

---

[6]Ranstorp, 1997.

[7]Bodansky, 1993; Hedges, 1995.

[8]The Islamic Call Party had perhaps several thousand active supporters in Iraq and fighters based in Iran during the Iran-Iraq War. It split, however, into several factions, some of which are in Iran and one of which is in Damascus.

Hezbollah actively spread propaganda against the Al Khalifa, but it was not linked to any actual acts of violence or to the larger demonstrations that occurred.[9]

Tehran's leaders are increasingly pragmatic in their expectations and use of local proxies. In the early 1980s, Iran used political violence in the Gulf to stir up unrest and to punish Iraq's allies, even though such political violence usually backfired, hardening Gulf state opposition to the clerical regime. Today, Iran increasingly uses proxies to influence government decisionmaking rather than to threaten a rival regime's very existence. In Lebanon, for example, Iran has supported the Lebanese Hezbollah's entrance into Lebanese politics. Subversion thus complements diplomacy. Regimes with pro-Iranian Shi'a movements in their countries know that if their foreign policies do not align with Tehran's, they may face domestic unrest. That is, Tehran may press local sympathizers to agitate on Iran's behalf. Iran also recognizes that local proxies offer them long-term influence.

Assassination: Targeting Dissidents

Iran uses political violence to attack critics of the clerical regime, particularly anti-regime Iranians. To this end, Iran has regularly assassinated members of the Mujahedin-e Khalq and Kurdish dissident groups.[10] It has also killed former members of the Shah's government and members of subsequent governments who became oppositionists, such as former Prime Minister Shapur Bakhtiar. Many of these assassinations were carried out by personnel affiliated with the Iranian Embassy or with Iranian-sponsored cultural and student organizations.[11] However, regime opponents do not regularly seek a haven in the Gulf states.

---

[9]"Bahrain: Defendants' Confessions," 1996; "Bahrain: Interior Ministry on Arrest," 1996.

[10]The Mujahedin-e Khalq is a radical group that endorses a synthesis of militant Islam and Marxism. In the past, it has supported the killing of U.S. servicemen and the takeover of the U.S. Embassy in Tehran. The movement was placed on the official State Department list of terrorist groups in 1997.

[11]Amos, 1994; O'Ballance, 1997; and Hoffman, 1990, p. 20.

Pressing Rival Governments

Political violence also enables Iran to press area governments to change their foreign policies. The Iranian-supported Da'wa group, which originated in Iraq but became affiliated with what later became the Lebanese Hezbollah, carried out six bombing attacks in Kuwait in 1983, with personnel, weapons, and explosives smuggled from Iran. Throughout the mid-1980s, Iranian-backed groups attacked U.S., French, Kuwaiti, Jordanian, and other targets associated with perceived backers of Iraq, to dissuade these governments from supporting Baghdad.[12] Iran also has used political violence to discredit the Saudi regime. Throughout the 1980s, Iran orchestrated demonstrations at the *hajj* that spilled over into violence; in 1987, hundreds of Iranian pilgrims died in riots in Mecca.[13] In 1989, a Hezbollah offshoot planted a bomb in Mecca that killed one person. The goal of such attacks was to punish the Kingdom for its support of Iraq and its ties to Washington.

Vanguard of Islamic Revolution: Enhancing the Role of Ideology

Although Iran does not hesitate to use political violence abroad to promote its state's objectives, it is also motivated to use violence to promote ideology, to act as the "citadel of Islam." To assert its support and leadership of the radical Islamist cause, Iran tries to maintain contacts with many radical Islamist groups of all stripes.[14] Iranian-backed terrorists attack secular intellectuals in the Muslim world to demonstrate Iran's commitment to spreading the Islamic revolution. The terrorist attacks on translators of Salman Rushdie's book *Satanic Verses* in Italy, Norway, and Japan—attacks inspired by, if not necessarily directed by, Iran—are an example of this tendency.

---

[12]Amos, 1994. Later, freeing the terrorists captured in Kuwait became a major goal of terrorists in Lebanon and in Kuwait.

[13]Iran also uses its local contacts and Iranian nationals to monitor U.S. forces and deployments.

[14]Iran has hosted radical conferences to which it invited representatives of leading Palestinian and Lebanese terrorist groups to discuss operations against Israel. Moreover, it maintains contacts with Islamist groups from the Gulf, the Levant, the Maghreb, and other parts of the Muslim world.

Similarly, Iran supported attacks against the United States by groups such as the Lebanese Hezbollah in the early 1980s to demonstrate its commitment to resisting U.S. hegemony. Iran, however, is able to control its use of political violence, employing it when the leadership sees it as beneficial and restraining itself when the time is not opportune.[15]

The fervor of the heady days immediately after the revolution has diminished. Even many former radicals now recognize that the complete Iranian model is not likely to be emulated, even by sympathetic groups such as Lebanese Hezbollah. The policy of spreading Iran's credo has broad support among Iran's leadership, but the specifics of the policy are hotly debated. This debate has two dimensions: tactics and priorities. Even relative moderates believe that Tehran should not abandon its export of the revolution, because Iran would lose its special character. Rather, the question is *how* the exporting should be done and what priority it should take beside other regime goals. So far, Iran's new President Mohammed Khatami has engaged in a "charm offensive" in his relations with the Gulf states, courting Gulf leaders while playing down Iran's desire for regional hegemony. Many Iranian leaders support this gentler approach to spreading the message of the Islamic revolution. Yet almost none of Iran's leaders are willing to give up their support for the export of the revolution in its gentler forms of propaganda and support for sympathetic Islamist activists.

Decreased Iranian Support?

Gauging the degree of Iranian support for political violence is difficult. We perceive that overall levels appear to be dropping steadily. Iranian propaganda on behalf of radicals or against Gulf leaders has cooled considerably. Tehran no longer decries Gulf leaders as corrupt and un-Islamic, as it regularly did in the 1980s. Levels of Iranian support can also be gleaned from the Gulf states' reaction to Tehran. Since 1997, rhetoric has cooled and relations have warmed. Perhaps most important, enthusiasm for Iran's message among the population of the Gulf appears to have waned. The continuing

---

[15]Green, 1995.

economic and political problems in Iran are well known in the Gulf. Moreover, since the death of Ayatollah Khomeini, no religious leader directly tied to the Iranian regime commands widespread support in the Gulf.

IRAQ'S USE OF POLITICAL VIOLENCE

Iraq has been less active than Iran in its use of political violence in the Gulf, for a variety of reasons. Iraq engaged in little or no political violence in the Gulf during the 1980s, primarily because the Gulf states were its allies in its war against Iran. After the Iraqi invasion of Kuwait in 1990, area governments removed or shrank from the official Iraqi presence, greatly reducing Baghdad's ability to carry out terrorist attacks.[16] In the future, Saddam is likely to back oppositionists throughout the Gulf if he can. The 1993 attempted assassination of President Bush suggests that Iraqi intelligence still can conduct limited operations and will use them for vengeance. But we have no evidence that Saddam currently is plotting to conduct terrorist attacks in the Gulf.

As does Iran, Iraq uses political violence to harass opposition figures. Iraqi intelligence agents have killed former regime officials, defecting diplomats, and other dissidents.[17] Iraq, however, is far less sophisticated in its use of political violence than is Iran. Thus, Baghdad is less able to press foreign regimes, create local proxies, or use violence to gain ideological credibility. In general, Baghdad has used its Foreign Ministry and Iraqi Airways to provide cover for its intelligence operations rather than operating through proxies.[18] During Desert Storm, the United States and its allies shut down much of Iraq's terrorist capabilities by expelling Iraqi diplomats and surveilling facilities associated with Iraq—obvious sites for any government to monitor.[19]

Baghdad also is willing to use groups for hire if its own intelligence services cannot carry out the attacks on their own. In the past, it

---

[16]"Saddam's Spies," 1995.

[17]"Saddam's Spies," 1995.

[18]"Saddam's Spies," 1995.

[19]Tucker, 1997; authors' interviews.

worked with the Abu Nidal Organization, the Palestine Liberation Front, and the Arab Liberation Front[20]—groups that have a veneer of ideological solidarity but that cooperate with Iraq primarily for the money it offers as a quid pro quo.

Unlike Iran, Iraq has generally not employed political violence and subversion as integral parts of its overall foreign policy. Iraq's leadership—its rhetoric to the contrary—does not truly consider itself to have a special ideological mission in the Arab or Muslim world and thus is less eager to proselytize. In addition, Iraq's nominal Baathist ideology, which combines elements of pan-Arabism and socialism, has largely been discredited in the Arab world since the 1960s.[21] As a result, Iraq has few fellow travelers in the Arab or Muslim world. Finally, Saddam Husayn, in contrast to Iran's leaders, has a short-term foreign policy focus. Consequently, Iraq is generally unwilling to invest resources in building local proxies whose influence may take years to develop.

Somewhat surprisingly in view of the brutal nature of his regime, Saddam has avoided fomenting mass terror abroad. In general, Iraq has not sponsored car-bomb attacks or insurgencies. Saddam prefers to kill individual leaders who offend him, rather than trying to stir up widespread popular unrest.

NONSTATE ACTORS

In addition to individual states, various radical and religious groups active elsewhere in the Middle East have periodically manifested a presence in the Gulf. The Lebanese Hezbollah and the Muslim Brotherhood in particular have influenced Gulf politics. Hezbollah has directly supported Gulf radicals; the Muslim Brotherhood's influence has been more ideological, providing an intellectual justification for anti-regime and anti-Western ideas.

The Lebanese Hezbollah often has acted as Iran's proxy in the Gulf, arming and training Gulf dissidents who acted on Iran's behalf.

---

[20]"Saddam's Spies," 1995.

[21]Pan-Arabism has failed to bring the Arab world together or even to promote harmony in the region. However, it does retain a degree of popular appeal despite its many failures.

Many Gulf Shi'a regularly visit Lebanon, and Hezbollah clerics from Lebanon are seen by some Gulf Shi'a as teachers whose instruction should be followed. Hezbollah also offers an excellent example of how to exploit local religious networks to create a strong political movement. Hezbollah grew as a movement partly by exploiting traditional religious hierarchies in Lebanon. It married these religious networks to a political agenda, creating an organization capable of concerted, directed operations as well as local initiative.

Although the Muslim Brotherhood is quite influential in the Gulf, it does not appear to have directly supported any radical activity. Muslim Brotherhood theology, like that of Hezbollah, calls for a state where Islam is the sole source of law. Some Muslim Brotherhood thinkers endorse the use of violence against Muslim rulers who are not sufficiently zealous. However, unlike Hezbollah, the Muslim Brotherhoods of Egypt and Jordan do not call for the overthrow of their respective governments by force, and neither group is well armed.[22] Many Egyptian and Jordanian workers in the Gulf are members of the Brotherhood and have disseminated their beliefs among Gulf residents: The Islamic Constitutional Movement in Kuwait is one example of a group whose doctrine is heavily influenced by the Brotherhood's teachings. Since many of these Egyptian and Jordanian workers are teachers, and often teachers of religion, the Muslim Brotherhood exerts a particularly strong influence.

---

[22]Before the early 1980s, the Syrian Muslim Brotherhood had perhaps thousands of militants under arms. After a brutal crackdown by the Asad regime in the early 1980s, its revolt died out. The Egyptian Muslim Brotherhood, which once violently resisted the government, began calling for peaceful change and working with the system in the 1980s. More-radical groups such as the Islamic Group still use violence against the government.

Chapter Five

GULF GOVERNMENT STRATEGIES FOR COUNTERING POLITICAL VIOLENCE

Effective government action can often avert political violence. Even in countries that are poor, ethnically divided, and beset by aggressive neighbors, a strong, resourceful regime can maintain peace. Governments can coerce or bribe opponents, offer opposition figures a voice in decisionmaking (or at least appear to), placate or threaten foreign powers, and otherwise manipulate the political equation to favor stability over unrest.

Despite the many problems besetting the Gulf regimes and their turbulent environs, regimes generally have kept the peace in their countries.[1] Gulf ruling families have proven skilled at anticipating—and preventing—political violence before it has become widespread. The regimes do so by employing a mix of sticks and carrots, using aggressive security services to monitor, and at times suppress, opposition while co-opting potential opposition leaders with financial incentives, jobs, and high-status positions. In addition, regime leaders have changed their outside appearance to match the issues of the day while maintaining their hold on power. Regime strategies shape politicization in a nonviolent manner and inhibit opposition organi-

---

[1]Monarchy in the Gulf is in many ways a different institution from monarchy in the West. Except for Saudi Arabia, the Gulf monarchies were instruments of British power for many years and thus did not have to construct their own apparatus for rule; regardless of their merit or the wishes of their followers, they stayed in power because they pleased the colonial government. This background, along with an influx of oil wealth, has freed the monarchs to make social bargains with key interest groups. Anderson, 1991, pp. 2–14.

zation; they do not affect the atmosphere of discontent. Thus, these tools may prevent widespread instability but will not stop all anti-regime or anti–U.S. violence.

This chapter examines the Gulf states' strategies to prevent political violence. These strategies explain why, despite the widespread grievances and regular interference of foreign states, the Gulf regimes face only a limited threat from political violence. The chapter describes tools ruling families use to manage dissent and notes how these tools affect anti-regime organization and the various violent politicizing factors identified in Chapter Three. After this general discussion, it evaluates the strengths and weaknesses of each of the four Gulf governments, as well as the particular problems each faces and the preferred strategies for ameliorating them.

STRATEGIES TO MANAGE DISSENT

Recall that Chapter Three concluded that violent individuals often are motivated by political alienation, defense of traditions, societies that glorify violent individuals, and outside interference. Moreover, to be truly effective, opposition forces also need to be well organized.

To offset these factors, the Gulf states rely on a combination of six tools to counter political violence:

- Maintaining strong security services to suppress dissent
- Co-opting regime opponents
- Dividing potential critics
- Seizing on potential anti-government issues as their own
- Making nominal efforts to include the citizenry in decision-making
- Having an accommodative foreign policy to appease foreign aggressors and seize the moral and ideological high ground.

Together, these strategies hinder anti-government organization, lessen hostility toward the regime, and otherwise reduce the immediate potential for political violence.

Maintain Strong Security Services

Security forces in the Gulf do not hesitate to suppress dissent.[2] The Saudi services closely monitor all organization—political, religious, and otherwise—in the Kingdom.   Bahrain's police force and the Bahrain Security and Intelligence Service (BSIS) have arrested and jailed participants in anti-regime demonstrations, and they suppress any gathering of protesters almost immediately.   The security services in Kuwait and UAE are less active, because there is only a low level of domestic opposition in those regimes; however, they vigilantly monitor the large expatriate worker populations in their countries and guard against foreign-backed political violence.[3]

Strong and efficient security services explain a large part of why most of the Gulf states experience fairly little unrest and only modest political opposition, despite having their share of disgruntled citizens. Perhaps most important, strong security services anticipate and suppress anti-government political organization.   Potential leaders fear to organize, recognizing that they will soon be discovered and possibly jailed or exiled.   Similarly, followers know that supporting opposition activity will hurt their chances of receiving a state job and may land them in jail.   Moreover, any public punishment would embarrass the activists' families: a potent deterrent in small countries where family ties are paramount.[4]

Coercion also shapes the Gulf's intellectual environment and offsets the corrosive influence of outside powers.   Security services in several Gulf states monitor intellectuals and spiritual leaders, leading both groups to avoid strong anti-government statements.   Thus, these potential critics become voices of restraint. All the Gulf regimes pay particular attention to foreign-inspired political activity in their countries.   Security services often monitor individuals who study or travel abroad, particularly those who travel to Iran or Lebanon, upon

---

[2]Regime security forces' ranks often are dominated and staffed by foreigners.   In Bahrain, for example, the security services were headed by a British expatriate until 1998.   The ranks are filled with Pakistanis, Indians, Baluchis, and others.   Similarly, other Gulf states recruit from foreign countries or rely on particularly loyal tribes or regions to staff the ranks of the services.

[3]Bahry, 1997; Cordesman, 1997a; authors' interviews.

[4]Authors' interviews.

their return. After the Khobar bombing, the Saudi government began scrutinizing the activities of Saudis who had fought in Afghanistan. Thus, foreign government agents—as well as many innocent citizens—are quickly rounded up if they encourage political activity.

However, *indiscriminate* use of security services can backfire, leading peaceful reformers to support violence. When peaceful tactics fail to move the government and any sort of opposition is prohibited, reformers are apt to lose hope in the political system. As a result, political alienation increases. Indeed, when an opposition organization is destroyed, its members, particularly its leaders, are often forced underground to avoid arrest, imprisonment, or worse. Once underground, they become more dependent on clandestine techniques to survive and may have to seek foreign assistance. This dependence may lead them to turn violent, either to extort money from hesitant supporters, to intimidate potential informants, or to keep the goodwill of a foreign sponsor.

Co-opt Potential Dissidents

Gulf regimes are expert at deploying largesse to silence critical voices. Critics of all sorts, both secular and religious, often are given jobs or government contracts in exchange for their support. On several occasions, a once-critical religious leader has received a lucrative position in exchange for his support or an academic critic has become the head of a government-sponsored institute.[5] Those who dissent jeopardize government patronage. For example, after opposition to the Al Khalifa grew in 1994 and 1995, the Bahraini government dismissed several critical professors and government employees from their jobs.[6]

All Gulf governments are remarkably skilled at using their control over their national economies to ensure their hold on power. In Kuwait and Saudi Arabia, perhaps 90 percent of citizens work for the government. In Bahrain and the UAE, large numbers of people hold government positions, often working simultaneously in family businesses. The Al Sabah work closely with wealthy Shi'a families and

---

[5]For examples of co-optation, see "Your Right to Know," 1995.

[6]Dunn, 1995; Cordesman, 1997a; authors' interviews.

use enormous financial resources to gain those families' support or at least to avoid their opposition. Even the Saudi Shi'a are included. Despite rampant discrimination against Saudi Shi'a in general, the Al Saud provided that community with additional funding after demonstrations in 1979. The ruling families also exercise more-subtle forms of financial control, often providing housing, health care, and other important benefits and thereby giving the regimes even more leverage over their citizens.[7]

To control the media, Gulf governments rely on subsidies and the threat of suspending publication rather than on formal censorship. Gulf governments often pay editors and reporters directly and provide funding for publication—all conditional on laudatory coverage of government activities and little coverage of opposition. Most journalists in the Gulf are expatriates from other Arab countries. Thus, they have little status and are entirely dependent on the goodwill of the state to stay in the country. When deemed necessary, Gulf governments will suspend publication.[8]

By co-opting critics, Gulf governments alleviate much of the immediate social tension. Potential critics' aspirations, for example, can be fulfilled on an individual level, with many disaffected leaders receiving a subsidy, official position, or other token of wealth and esteem. The regimes build religious centers, medical facilities, and other services to placate disaffected areas, using the promise of assistance to reduce anger.[9]

Co-opting opponents also helps regimes counter violence stemming from political alienation and the intellectual environment. The tokenism of including individuals from different social groups in government and other high-status positions demonstrates that the community in question is not completely excluded from power, thus reducing somewhat their community's overall dissatisfaction. When intellectuals and religious leaders are co-opted, they are less likely to call for anti-regime violence.

---

[7]Ghabra, 1997; authors' interviews.

[8]Cordesman, 1997a.

[9]Zonis, 1971.

Co-optation can also interfere with anti-regime organization. By providing would-be leaders—as well as the media and other voices that might publicize problems instead of downplaying them—with incentives to support the regime, it can reduce anti-government sentiment.

Divide and Rule

Gulf governments also are adept at creating divisions within communities and at fragmenting political opposition. To win over Shi'a elites, the government of Bahrain expanded its appointed advisory council in 1996, giving it a Shi'a majority. This, combined with regular government largesse to certain families, has led many wealthy Shi'a to support the government and work with it against the poorer radicals. Saudi Arabia and Kuwait have long collaborated with Islamic forces against leftist Kuwaiti groups in their countries. In the 1970s, for example, the Al Sabah supported the Social Reform Society, a then-nonpolitical Islamic group, against Arab nationalist groups.[10]

Divide-and-rule tactics often reduce elite aspirations, because the regimes portray themselves as the best available alternative to popular rule. The Al Khalifa in Bahrain, for example, are past masters at exploiting Sunni suspicion of Shi'as. Even Sunnis who are appalled by the Al Khalifa, and favor a return of the National Assembly, have gradually withdrawn support from the reform movement, fearing that the Shi'a will dominate it. The Al Khalifa play up this division. For example, in 1995, they arrested Shi'a while permitting many Sunni activists to remain free. The Al Khalifa also divided the Shi'a community, co-opting wealthier Shi'a while cracking down on poorer ones. Thus, Bahrain's opposition is rent by both sectarian and class divisions.[11]

Divide and rule also hinders anti-regime organization. Dividing groups reduces their overall size, making them less influential and less dangerous. Even more important, divided groups quarrel among themselves, diverting attention from anti-regime campaign activity.

---

[10]Ghabra, 1997; authors' interviews.

[11]Bahry, 1997; authors' interviews.

Exercise Ideological Flexibility

Despite the traditional nature of monarchical political systems, Gulf ruling families are able to bend with the prevailing political winds. During the 1950s and 1960s, they often claimed to champion Arab nationalism, offering token support in the fight against Israel and funding revolutionary Palestinian groups.[12] After the 1979 Iranian revolution, the Gulf leaders changed coloration, portraying themselves as pious Muslims, fervent in their support for traditional religion.

Under the guise of providing ideological support, the ruling families have tightened their hold on power by replacing local, autonomous institutions with state-sponsored ones. Claiming only that they sought to support religion, Gulf leaders have in fact co-opted many religious leaders and undermined their independent bases of support. Shi'a, long excluded from patronage by discrimination, also had an independent base of support. In Bahrain, the Al Khalifa slowly recognized that this autonomy posed a challenge; in April 1996, they set up the Islamic High Council to screen religious leaders and provide religious guidance. Ostensibly, the council aided religious activities by providing additional funding, but the council also ensured that anti-government religious activity did not occur.[13]

Such ideological and practical measures counteract a tremendous amount of immediate hostility on the part of Gulf residents. Alienation, both moral and political, is reduced by the Gulf leaders' public identification with the *zeitgeist*. On a practical level, government measures in the name of the cause exceed any incentives offered by opposition groups. A government-run "Islamic" clinic, after all, will be more lavish than a private Islamic clinic will be.[14]

---

[12]In 1967, Riyadh sent a brigade to Jordan, but it moved so slowly, and the Israelis moved so quickly, that it did not arrive until the war was over. In 1973, the Saudis sent a brigade that was not supposed to engage in conflict; however, when Israel broke through the Syrian lines, it encountered the Saudi forces and engaged in a minor skirmish.

[13]Cordesman, 1997a.

[14]Perhaps surprisingly, the Gulf leaders' suppleness also allows them to stand as tradition's defenders. When in doubt, the Gulf leaders move slowly and work closely with the country's religious and tribal establishments. These institutions depend

Ideological flexibility also offsets outside interference. After the Iranian revolution, for example, Khomeini and other Iranian clerics lambasted the Gulf monarchs as un-Islamic and corrupt. In response, the Gulf royal families made public shows of their piety. The Al Saud monarch even adopted the title "Custodian of the Two Holy Places" (the sacred sites of Mecca and Medina, both of which are in Saudi territory) to bolster his credentials.

Yet, ideological flexibility can backfire, especially when it lends legitimacy to violent individuals. For example, the Gulf states sought to burnish their Arab nationalist credentials by embracing radical Palestinians. They implicitly (and at times quite publicly) supported the violence used to advance this cause. By the same token, Gulf state support for the radical Islamic movement HAMAS in the West Bank and Gaza, anti-Soviet fighters in Afghanistan, and anti-Israeli operations in general, has provided these warriors for Islam with the imprimatur of legitimacy.

Encourage Pseudo-Participation

To varying degrees, the Gulf states also use appointed and representative institutions to provide a forum for discussion and input into decisionmaking. Where these institutions are more than window dressing, such as in Kuwait, they demonstrate that the regime is accessible to the people and reduce the sense of political alienation created by the ruling family's domination of politics. Even where they are weak, they suggest that the ruling families are willing to go outside their own ranks when weighing decisions.[15]

In Bahrain, the late Amir Isa gradually extended the role of an appointed council in response to continued unrest. In the fall of 1992, Isa appointed a 30-member Consultative Council in response to post–Desert Storm calls for a greater popular voice in decisionmaking. Initially, the council was evenly split between Shi'as and Sunnis. It had no legislative power, and its initial meetings were not reported in the media. In 1996, after two years of anti-regime protests, the

---

heavily on the ruling families, and a threat to the regime could undermine their influence.

[15]Authors' interviews.

Amir expanded the size of the council, appointing more Shi'a members. He also increased media coverage of council events.[16]

Pseudo-legislative fora are particularly weak in Saudi Arabia and the UAE. After calls for reform became increasingly significant following the Gulf War, Saudi's King Fahd announced in March 1992 that he would appoint a consultative council; in August 1993, he chose 60 members to serve on it. Council members represent a cross section of the Saudi elite, including religious officials, merchants, university professors, and technocrats. The UAE has a Federal National Council, whose members are appointed by the emirate rulers. The council does engage in some debate over government policy, such as over the allocation of services to various emirates.[17]

Offsetting the weakness of these institutions is the small size of Kuwait, Bahrain, and the UAE. To varying degrees, all Gulf ruling families and elites offer access to their citizens by holding regular, but informal, meetings wherein citizens can air their complaints, petition for redress of grievances, or otherwise try to influence local and national politics. As one Bahraini interlocutor noted, "I don't worry too much about whether I have a vote or not—after all, I can talk to someone who talks to the ruling family simply by picking up the phone."[18] By attending local gatherings and simply keeping their doors open, ruling families generally have access to popular opinion.

This sense of inclusion is particularly useful in anticipating and deferring violence generated by political alienation or frustrated elite aspirations. The local gatherings, informal talks, and weak legislatures bolster regime claims that they respect, and listen to, the voices of the citizenry. Indeed, the one-to-one contact with the ruling families generates a sense of common identity between the rulers and the ruled. Elites in general have more access to the ruling families than do ordinary citizens and are often chosen to sit on local or national councils. Thus, their resentment of ruling families is lessened somewhat by the higher status accorded them. Furthermore,

---

[16]Cordesman, 1997a; authors' interviews.

[17]Gause, 1994; Rugh, 1997.

[18]Authors' interviews.

the lack of a regional government that does allow widespread participation limits the demand for more pluralism in the Gulf.

Engage in Accommodative Diplomacy

The Gulf states try to placate potential foreign adversaries with low-
profile foreign policies and generous aid. In the 1960s and 1970s, the
Gulf states funded radical Palestinian groups and "front-line"
states—Syria, Egypt, and Jordan—in their fight against Israel, in an
effort to insulate themselves from criticisms that they did little to advance the Arab nation's cause. Indeed, they initiated the oil embargo
of 1973 to counteract criticism that they were not committed to Arab
nationalism. Similarly, when political Islam grew in importance, the
Gulf states aided some radical Islamist groups in order to preempt
criticism.[19]

This conciliatory approach has met with mixed success since the
1950s. Egyptian President Nasser, for example, at times reduced his
criticism of the Al Saud in response to Saudi blandishments, but he
did not hesitate to attack them vociferously when his ambitions or
domestic political position required it. In addition, as Kuwait
learned so tragically, years of support and subventions can whet the
appetite of an aggressor rather than sate it. Most important, it is
difficult to embrace the messenger while rejecting the message.
Appeasing Nasser required some domestic toleration of Arab
nationalists; satisfying future aggressors may also require allowing
the spread of potentially subversive ideas.

The above strategies overlap in practice. For example, efforts to co-
opt leaders also involve attempts to divide and rule the population.
Ideological flexibility at times requires the nominal inclusion of os-

---

[19]Yet the challenge of political Islam championed by Iran exposed a tension in the
Gulf states' diplomatic strategy. During the Iran-Iraq War, the Gulf states faced a no-
win situation: On the one hand, if they failed to aid Iraq, the Baath regime would
renounce the Gulf states and stir up Arab nationalism against them; alienating Iran, on
the other hand, would lead to continued Iranian attempts to inspire revolution on the
peninsula. Iran's version of political Islam favored Shi'ism; consequently, it received
less sympathy in the Gulf than it might have had it been Sunni-dominated. Moreover,
Tehran regularly sponsored subversives in the Gulf and seemed unmoved by the Gulf
regimes' calls for peaceful relations. Thus, the Gulf states confronted Iran openly (by
their standards) by providing billions of dollars in aid to Iraq each year.

tensible opposition figures. Moreover, the Gulf leaders do not consider the strategies as separate, but use them together to manage dissent.

IMPACT ON POLITICAL VIOLENCE

The above strategies do not strongly affect the atmosphere of discontent discussed in Chapter Two: They do not stop social modernization, revive stagnant Gulf economies, or reduce corruption.[20] Instead, they are short-term palliatives. Regime strategies do hinder opposition organization, which is necessary for effective political action, and they mitigate the politicizing events that often lead disaffected individuals to become violent.

Stopping Anti-Regime Organization

Almost all the strategies listed above reduce the ability of opposition groups, both violent and peaceful, to organize. Leaders who seek to rally their followers against the regime are often both brutalized and bribed, facing the choice of years of imprisonment for continued opposition or a lucrative position if they desist. Their followers, to a lesser extent, face a similar choice between harassment and reward. Generally, only the most dedicated continue in their opposition.

Pseudo-participation, divide-and-rule, and ideological-flexibility tactics also reduce anti-regime opposition. Gulf rulers are adept at playing off their rivals, tempting them with promises of more support or recognition and generally shifting the issue from working together against the government to working with the government against each other. Similarly, rulers often take the wind out of the opposition's sails, announcing (but not necessarily implementing) reforms along the lines demanded by regime critics. For example, both the Saudi and Bahraini "parliaments" were created in response to calls for greater regime openness and responsibility. Such moves have

---

[20]Nevertheless, accommodative diplomacy does reduce foreign-organized violence somewhat and has sheltered the Gulf states from substantial unrest. In Kuwait, the National Assembly has somewhat countered the perception that the Al Sabah exclude others from decisionmaking.

satisfied some critics, particularly those invited to participate in the parliaments, without substantially reducing regime authority.

A strong, well-organized movement is not necessary for political violence. However, it is generally essential for the most politically effective types of violence. To sustain an anti-government campaign, advance the cause of a political movement, or otherwise use violence as part of a long-term agenda, the group in question must have established leaders, a *modus operandi*, an intelligence network, a recruitment arm, sources of money, and perhaps even links to outside supporters. When government policies hinder organization, these requirements are difficult to fulfill.

Organization is particularly important if groups are going to act as another state or movement's proxy. In such cases, well-timed operations are essential if the operational or political effect of the local group's action is to serve the particular needs of the outside group. In addition, the coordination and control mechanisms must be advanced to ensure that the local group remains subordinate.

But unorganized groups can still use violence, which can have a political effect. Assassinations, riots, and even attacks on government facilities are all within the range of radical individuals and small groups. For example, the November 1995 bombing of the Office of the Program Manager for the Saudi National Guard (OPM/SANG) that killed five Americans appears to have been carried out by a small group of Saudis acting on their own, with only limited support and training from outside groups.

As Figure 5.1 suggests, regime strategies have a tremendous impact on anti-regime organization. Gulf security services in particular play an important role in preventing organization. In addition to harsh security services, an array of gentler tactics such as co-optation and encouraging pseudo-participation also limit organization.

Preventing Dangerous Politicization

The strategies used by Gulf regimes to preserve power have important, and different, effects on the various politicizing factors, as noted in Figure 5.2. Some politicizing factors—such as the plethora of violent heroes in the Muslim and Arab world—are difficult for the

RAND*MR1021-5.1*

Impact on Opposition Groups	Strategies					
	Maintain security services	Co-opt	Exercise ideological flexibility	Divide and rule	Encourage pseudo-participation	Engage in accommodative diplomacy
Leaders unwilling to lead	Significant	Significant	Moderate	Moderate	Moderate	Little
Followers unwilling to support leaders	Significant	Moderate	Moderate	Moderate	Moderate	Little
Foreign-power role minimized	Significant	Little	Moderate	Little	Little	Moderate

Little Impact ◯   Moderate Impact ⊜   Significant Impact ⬤

Figure 5.1—Impact of Gulf Strategies on Opposition Organization

RAND*MR1021-5.2*

Impact on Politicizing Events	Strategies					
	Maintain security services	Co-opt	Exercise ideological flexibility	Divide and rule	Encourage pseudo-participation	Engage in accommodative diplomacy
Political alienation	Little	Moderate	Moderate	Little	Moderate	Little
Outside interference	Significant	Little	Moderate	Little	Little	Moderate
Defense of traditions	Little	Little	Moderate	Little	Little	Little
Glorification of violence	Moderate	Moderate	Little	Little	Little	Little

Little Impact ◯   Moderate Impact ⊜   Significant Impact ⬤

Figure 5.2—Impact of Regime Strategies on Politicizing Events

governments to reduce. Yet, the various strategies used by Gulf regimes greatly mitigate many sources of violence, particularly political alienation and outside meddling.

And some regime strategies make certain factors worse at the same time that they help alleviate various problems. For example, the strong security services play an important—indeed vital—role in reducing unrest from outside meddling and keeping political opposition weak. Yet they often lead otherwise peaceful individuals to go underground and even to take up arms out of necessity. Similarly, ideological flexibility—generally an extremely effective strategy— makes it hard for the Gulf leaders to distance themselves from radicals, who later inspire others to use violence.

VARIATION BY COUNTRY

So far, we have discussed the Gulf regimes as facing common problems and using similar responses; in fact, they often differ markedly. This section examines how each Gulf government perceives the problem of political violence and discusses its preferred responses.

Saudi Arabia

Long focusing most of its energies on keeping elites satisfied and guarding against Shi'a Muslim unrest, the Saudi government has focused more attention since the end of the Gulf War on tracking foreign-backed radicals and on monitoring the activities of Sunni opposition movements, such as those discussed in Chapter Two and in the following subsection. When facing Shi'a unrest, or opposition groups that the government believes are tied to Iran, the Al Saud do not hesitate to suppress: In the late 1970s and early 1980s, the Al Saud arrested or detained hundreds of Saudi Shi'a, fearing that they were in league with Iran. The Kingdom has a range of intelligence and police forces that keep a watch on one another as well as on the population at large.[21] They regularly jail or exile political organizers and try to prevent even informal organization among the Shi'a.

---

[21]Saudi Arabia has a variety of security services. The 10,000-man Frontier Force monitors the Kingdom's borders. The National Guard, relying primarily on loyal tribes, also provides internal security. The national police force has more than 15,000

To keep Saudi elites happy, the Al Saud have long relied on buying their allegiance. Lucrative contracts and government positions (which ensure access to wealth) are used to keep leading families and tribes content. The Saudis also press elites who resist the government, initially threatening to cut off the flow of wealth to them and their families before adopting more coercive tactics.

Saudi repression is limited to anti-regime political activities. If citizens—excepting the Kingdom's Shi'a population—play by the rules, the regime does not restrict their activities. Economically, individuals have tremendous freedom. Even the level of repression can be surprisingly limited. The government carefully monitors potential dissidents, but it seldom beats or imprisons them, preferring instead to bribe them or their families or otherwise induce them to conform.[22]

The Al Saud are less likely than other area regimes to use pseudo-participation and ideological flexibility to reduce dissent. The Saudi ruling family is committed to a strong religious identity and, though it has supported Arab nationalists in the past, is often unwilling to change identities capriciously. And, in contrast to Kuwait and Bahrain, the Saudi middle class is new; there is no tradition of formal representative bodies in the Kingdom. As a result, the Saudi opposition agitates less for a more representative system than do regime critics in other Gulf countries.

The Saudi government also retains a tight hold over information in and about the Kingdom. Government data are almost nonexistent, and the regime discourages any outside reporting on events in the Kingdom. Saudi newspapers are easily the least informative of all the Gulf states' media. Even Saudi-owned newspapers published in London or elsewhere in the Arab world, while often outstanding in their discussions of politics elsewhere in the Arab world, seldom if ever cover the Kingdom.

---

men. Other important organizations include the General Directorate of Investigation and the Special Security Force, both of which are controlled by the Ministry of Interior. Some security tasks are also performed by the General Intelligence Directorate and the large special investigations forces.

[22]Authors' interviews.

Shi'a radicalism poses less of a threat to the Kingdom's stability than it did in the early 1980s, when Saudi Shi'a—inspired by the example of Iran's Shi'a revolution—repeatedly clashed with Saudi security forces. Since the 1980s, Iran's revolutionary fervor has abated, the Shi'a leaders recognizing that the Iranian model will not work in the Kingdom. Moreover, regime spending on the Shi'a community and co-optation of Shi'a leaders have dampened enthusiasm for confronting the government, so that few Shi'a today appear eager to embrace violence or support anti-government opposition groups.[23] At this writing, this diminution in conflict is fueled by a growing rapprochement between Saudi Arabia and the Islamic Republic of Iran.

The Enigma of Sunni Radicalism

The Al Saud base their legitimacy largely on being defenders of the ultra-orthodox Wahhabi school of Islam, which means that the many demands of radical Islamists elsewhere in the Muslim world have already been met in Saudi Arabia: Islamic law is Saudi law; women face severe restrictions on their participation in any activity outside the home; and the regime enforces public observance of religion.

Yet despite this support for public piety, Sunni radicals regularly criticize the regime as un-Islamic.[24] These radicals oppose the very concept of secular authority and are zealous in their condemnation of any deviation from their view of the true faith. Regime corruption, ties to the United States, and charges that the royal family is not providing for its citizens are common complaints.[25]

The Sunni radical challenge is not new, and the regime is well aware of its serious dimensions. Even before the founding of the Saudi state, ultra-conservative Saudis found fault with the Al Saud. Periodic criticism occurred as the Al Saud consolidated power. Clashes at times turned violent, with the regime using the army against radi-

---

[23]Gause, 1994; authors' interviews.

[24]Even nonviolent religious leaders often oppose the regime. In 1992, 107 Saudi religious leaders signed a petition that called on the government to implement, among other things, Islamic law more strictly, reduce corruption, and sever relations with non-Islamic countries and the West. It also called for religious leaders to have a formal role in government. Cordesman, 1997b.

[25]Cordesman, 1997b.

cals. In November 1979, Sunni radicals seized the Grand Mosque in Mecca during the *hajj*, claiming that the Al Saud was illegitimate because it transgressed against the puritanical Wahhabi credo. Security forces stormed the facility, leaving dozens dead.[26]

Radical complaints have grown in recent years, and they increasingly focus their grievances on the United States as well as the Al Saud. In 1991, Saudi radicals wounded two Americans in an attack on a shuttle bus in Jeddah. On November 13, 1995, Sunni radicals killed seven people, including five Americans, when they bombed the U.S. Army Materiel Command's Office of the OPM/SANG. The individuals arrested for the attacks, Saudi extremists who had fought in Afghanistan and Bosnia, sought to end the U.S. military presence in the Kingdom.[27]

Past attempts to divide and rule have sown the seeds for more-recent challenges, such as the emergence of Islamic radicalism in the 1970s and 1980s. Encouraging religious radicals to organize in the 1960s and 1970s, the regime correctly anticipated that doing so would reduce the influence of the then-dominant school of Arab nationalism, which was often anti-monarchist.[28] However, the regime strengthened the very groups that would later challenge it: groups of Sunni radicals enabled to create a network by the mosques and organizations the regime supported in the 1960s and 1970s, the most ardent with additional contacts as a result of their support for, or participation in, fighting in Afghanistan.

Today's Sunni radicals are strong and may be well organized, although supporting data are fragmentary and anecdotal at best. Both the regime's claim to rule in the name of Islam and its support for religious causes as a divide-and-rule tactic have made religious groups the strongest opposition movement in the Kingdom. The Al Saud face a dilemma: Continued organization might strengthen religious

---

[26]Gause, 1994.

[27]Cordesman, 1997b.

[28]Ironically, Israel used the same tactic to undermine the secular PLO and today is locked in a bitter conflict with Islamic groups it had originally helped to create. While the PLO found a *modus vivendi* with Israel, Hezbollah and Islamic Jihad have not. At times, the cure can be worse than the disease.

opposition, but a crackdown on religious groups will decrease the regime's strongest claim to legitimacy.

So far, the regime has relied primarily on co-opting Sunni radicals while demonstrating its own Islamic credentials. The Al Saud have tried to increase official and popular deference to Islamic law and have strengthened the religious police. Religious leaders also increasingly control the Saudi education system. However, as a result of these tactics, Sunni radicals are well entrenched in Saudi Arabia's religious institutions and economic structure.[29] Consequently, the security services are less able to control their activities. Some radical Sunni groups, particularly those from the Najd, may have ties to ruling-family members and to local elites. Often, the radicals are the brothers and schoolmates of government officials.[30]

However, since the end of the Gulf War in 1991, and particularly after the 1995 and 1996 bombings of U.S. facilities in the Kingdom, the regime has stepped up efforts to monitor, repress, or simply control religious extremists. The security services monitor religious leaders more now, and the King has dismissed religious leaders who would not condemn the 1992 petition-signers. In 1993, the government required Islamic groups that formerly aided radicals in Algeria, Sudan, Afghanistan, and elsewhere to obtain authorization before soliciting funds. The government has even arrested individuals suspected of supporting radicals both at home and abroad.[31]

Bahrain

The Al Khalifa fear Shi'a activism as the greatest threat to their rule, particularly when it is backed by Iran. In 1981, Bahraini Shi'a, backed by Iran, tried to seize power, and Iran subsequently funded anti-government Shi'a groups. As a result, the Al Khalifa carefully monitor the Shi'a community for unrest, and seek to keep it divided from its Sunni neighbors.

---

[29]Cordesman, 1997b.

[30]Authors' interviews.

[31]Cordesman, 1997b.

Bahrain also has the most competent security services in the Gulf. The Ministry of Interior is responsible for public security and uses the Public Security Force and the Bahrain Security and Intelligence Service to ensure internal order. The BSIS is very good at discriminating between its victims and others: It does not fire into crowds; and it does not lose control or repress indiscriminately. Most important, it has a superb domestic intelligence network. The BSIS also cooperates with other Gulf security services, and, in cooperation with the UAE, has scored some notable counter-terrorism successes, such as the detection and foiling of an Iranian-backed coup attempt in 1981. The country's small size makes the BSIS' task relatively easy.[32]

Bahrain has not hesitated to use its security services to clamp down on unrest, at times imposing collective punishment on villages in which protesters have been active. Bahraini opposition members claim that more than 10,000 people have been detained since 1994, and that the security services have injured more than 500 citizens and ransacked mosques and other religious gatherings.[33]

The Al Khalifa's response to unrest also has a foreign-support component. Bahrain is better able to co-opt potential oppositionists, and keep its economy afloat somewhat, with aid from its neighbors. This support, however, does not come without strings. In general, Bahrain's neighbors prefer that Bahrain repress its opposition rather than appease it with political and economic reforms.[34] Saudi Arabia in particular opposes concessions to Shi'a movements.[35]

Future stability is less certain. The British expatriate Ian Henderson, who had headed the BSIS since 1966, retired in 1998, and the competence of the security services after his departure is uncertain. Henderson focused the BSIS less on physical intimidation of the population-at-large and more on monitoring specific individuals and groups. Many observers believe that he prevented the BSIS from using more brutality and worked to limit the involvement of

---

[32]Authors' interviews; Cordesman, 1997a.

[33]"Bahrain Uprising: 3 Years Old," 1997.

[34]Shaykh Zayid, the leader of the UAE, has urged Bahraini leaders to reconcile with the Bahrain Freedom Movement.

[35]Authors' interviews.

Bahrain's armed forces in suppressing dissent. The competence of his successor, Khalid Bin Mohammed Al-Khalifa, is not known. Generally, however, the Al Khalifa are more likely to crack down than to conciliate. Absent Henderson, this repressive tendency may rise.[36]

The Al Khalifa's skills at winning over opponents and satisfying them with small rewards appear to be declining. Increasing numbers of the Sunni elite are disenchanted with the regime, and the Shi'a population in particular is hostile. Many of the ruling family openly scorn the Shi'a, limiting their ability to portray themselves as supporters of the Shi'a. Fear keeps much of the population in line today. During the recent violence, collective punishment may have exacerbated tension, creating support for the radicals. The government also reduced the number of Shi'a Muslims in top official positions and at the University of Bahrain.[37] The problems are heightened by Bahrain's relatively weak economic situation, which offers the regime fewer resources to bribe elites and satisfy the population.[38]

Kuwait

Kuwait's Al Sabah are probably the Gulf ruling family most skilled at co-opting their opponents and permitting a significant degree of genuine political participation. In contrast to other Gulf states, Kuwait has a nascent democracy with a functioning National Assembly. The assembly permits part of Kuwait's citizenry to participate in the public debate and to exercise limited control over decisionmaking.[39] The Al Sabah also included several opposition mem-

---

[36]Authors' interviews; Cordesman, 1997a.

[37]Jamri, 1997.

[38]In a major coup or a civil crisis, the Al Khalifa could turn to the Al Saud for assistance—although they would be loath to do so in all but the most dire circumstances. When the demonstrations in Bahrain began, the Saudi Arabian Interior Ministry declared that Bahraini security was inseparable from that of Saudi Arabia. When Shi'a unrest grew and riots began, the Saudi government ostentatiously sent a company of armored personnel carriers to Bahrain "for maneuvers." Although this unit did little, it did move around the island with great visibility, making it clear to all of Bahrain that Saudi Arabia was watching. Given that the Saudi government is well known for it brutal treatment of Shi'a in general, Bahrain's Shi'a community is probably hesitant to risk greater Saudi involvement. Authors' interviews.

[39]The franchise is extremely limited in Kuwait. Neither women nor expatriate residents can vote. For many years, only Kuwait men older than 21 who descend from

bers in the government. In addition to the National Assembly, the Al Sabah make an effort to include different segments of Kuwaiti society in their government; every Kuwaiti cabinet has a "Shi'a seat." The Al Sabah also call themselves a "ruling" rather than a "royal" family to emphasize the consensus that underlies their power, and Al Sabah members mix regularly with Kuwait's population at large.[40]

Kuwaiti intelligence, while competent, could take lessons from the BSIS. However, Kuwaiti intelligence has never faced the same internal challenge as has Bahrain's service, and thus has not had to develop the same ability to control and monitor the population in general.[41]

Kuwait's security service is primarily concerned with external enemies of the regime, particularly Iraq. The security service also monitors foreign workers, deporting many after a short time as a matter of policy. Workers from allies of Iraq during the war, such as Jordanians and Palestinians, have been greatly reduced in number. U.S. officials report that a top priority of the Kuwaiti security services is ensuring the security of the U.S. presence.[42]

Kuwait seems to have weathered the storm of Shi'a radicalism that followed the Iranian revolution and to have co-opted Sunni Islamist opponents. In 1985, Shi'a radicals attempted to assassinate the Kuwaiti Amir. Sunni radicalism appeared to be growing before the Gulf War. However, the Al Sabah's combination of co-optation and political liberalization has given these groups a greater voice in decisionmaking and increased their loyalty to the current political system.[43]

Ironically, the Iraqi threat has improved domestic stability by uniting Kuwaitis. Iraq's brutal six-month occupation and continued threats

---

families that resided in Kuwait in 1921 are eligible to vote—about 15 percent of Kuwait's adult population. Before the last election, the Al Sabah expanded the franchise to include the sons of naturalized Kuwaiti citizens, a decision that almost doubled the number of voters.

[40]Authors' interviews.

[41]Authors' interviews.

[42]Authors' interviews.

[43]Authors' interviews.

to Kuwait's security make many Kuwaitis highly security-conscious. They fear that airing their differences publicly might lead to foreign interference and believe that a united front is necessary. In addition, the invasion increased social solidarity, leading Kuwait's Sunni and Shi'a radicals to bridge their differences.[44]

The UAE

The UAE's vast wealth and small size have limited the scope of political opposition to date.[45] Only 19 percent of the UAE's citizens are Emirians, and through personal attention, UAE ruling families allow Emirians access to government. The federal nature of the national government also provides many points of access for citizens. Finally, the UAE's large oil reserves give the government tremendous resources to ensure the population's happiness and to buy off any dissent.

In general, the security services in the UAE are less intrusive than those of the other Gulf states, but this may be due to a lack of unrest in the federation. There are no political prisoners or exiles, and the regime does not extensively surveil citizens. To ensure security, each emirate has its own police force, and both the federal government— dominated by the emirate of Abu Dhabi—and the emirate of Dubai have their own internal security and intelligence organizations. The security services in the UAE keep a careful watch over the foreign community, including the relatively large Iranian national population. Many Iranians (and Emirians) are involved in smuggling between the two countries, and Dubai in particular is a conduit of goods from the West to Iran. In general, the UAE authorities look the other way.[46]

Figure 5.3 illustrates the varying strategies used by Gulf regimes. In general, all the Gulf regimes are particularly skilled at using their security services to prevent unrest and at co-opting potential dissidents. However, some governments, such as the UAE, have simply

---

[44]Authors' interviews.

[45]However, the lack of any discernible political opposition leaves open the question of how skilled the regime is at buying off, deflecting, or suppressing any unrest.

[46]Cordesman, 1997a; authors' interviews.

RAND*MR1021-5.3*

Country	Strategy					
	Maintain security services	Co-opt	Exercise ideological flexibility	Divide and rule	Encourage pseudo-participation	Have accommodative foreign policy (e.g., Iran and Iraq)
Saudi Arabia	⊜	⊜	⊜	◯	◯	⊜
Bahrain	●	⊜	◯	⊜	◯	⊜
Kuwait	⊜	●	⊜	◯	●	⊜
The UAE	⊜	●	⬤	⬤	⬤	⊜

Little Impact ◯   Moderate Impact ⊜   Significant Impact ●   Untested ⬤

Figure 5.3—Impact of Regime Strategy on Stability, by Country

encountered too little unrest for us to make a judgment on their ability to successfully accommodate and divide opposition figures.

CONCLUSIONS

Northern Persian Gulf area regimes employ an effective combination of sticks and carrots to prevent anti-government political movements from developing. Equally important, they cleverly use slogans of many of these movements as part of their overall ideological flexibility, dodging the bandwagon by riding on it. To reduce much of the immediate tension, particularly in Kuwait, governments also make small concessions in their domination of decisionmaking.

These strategies work primarily in two ways. First, they direct politicization, making it less likely that individuals will use violence in pursuing their political agendas. Second and equally important, they impede organization, making it difficult for individuals to use violence effectively or to act in large numbers. But, as noted in Chapter Two, the various strategies do not affect the fundamental grievances found in the Gulf. Thus, the potential for violence remains high.

The very effectiveness of the Gulf governments in fighting terrorism may lead to a growth in extra-regional terrorism against the United States and its Gulf allies. Osama bin-Ladin and other anti–U.S. Gulf radicals have announced that they will attack Americans anywhere. Because of the strong, and growing, U.S. force-protection measures in the Gulf, radical leaders must turn elsewhere in search of more-vulnerable targets. Thus, they train and support individuals who strike U.S. citizens and facilities in Africa, the Philippines, and the United States itself.

Chapter Six

IMPLICATIONS FOR THE UNITED STATES

The publicity given to the unrest and violence in the Gulf following the 1996 Khobar Towers deaths should not obscure the exceptional skill with which most Gulf regimes have contained and controlled internal dissent. For decades, the traditional monarchies weathered successive storms of Arab nationalism and Islamic radicalism. Gulf states should also be recognized for their ability to prevent significant dissent by the large and frequently resentful expatriate communities in their countries, comprising at least 37 percent of the total population in Bahrain and exceeding 80 percent of the population in the UAE. By monitoring these communities and expelling any individual who is even indirectly involved in anti-regime activity, the Gulf governments have prevented these large communities from being a potent political force.

Nevertheless, political violence remains a serious problem for the coming decades, both for the United States and for area regimes. In this final chapter, we look at ways in which political violence may increase in the coming years and discuss why potential solutions will solve only part of the problem at best. We next look at the trade-offs and dilemmas the United States and its Gulf allies will encounter when trying to fight political violence. We then emphasize how violence, in turn, may undermine U.S. domestic support for a military presence in the Gulf and impede cooperation between the United States and its allies. Because of this danger, the current emphasis on force protection should be continued in the coming years.

CLOUDS ON THE HORIZON

Political violence in the Gulf is likely to continue at limited levels in the coming years. Although techniques employed by Gulf governments may reduce anti-government sentiment while controlling hostile organization, the basic problems of corruption, stagnant economies, and unrepresentative governments remain. Indeed, several problems may increase in the coming years: diminishing government resources, growth in anti-regime groups' ability to organize overseas, and inadvertent tensions among the various strategies regimes use to fight violence.

Violence may increase as the resources available to governments diminish. As regime revenues fall, Gulf leaders become less able to buy off dissent. In Bahrain, the Al Khalifa already are facing difficult choices of whether to sate family greed or buy off other Bahraini elites. Over the years, as the Gulf populations grow and oil wealth stays constant or even diminishes, this problem will intensify.

More and more, anti-regime Gulf groups also are increasingly able to organize themselves overseas. So far not threatening the security of Gulf regimes, this organization does represent a chink in the Gulf states' armor. Overseas organization hinders the regimes' most effective measure against political unrest—suppressing organization.

As well, several of those very strategies the Gulf states use to combat political violence can, in the long term, actually contribute to anti-regime politicization and, eventually, political violence. In particular, the Gulf states' reliance on security services, especially in Bahrain, can lead to future problems. Because security services fight all forms of political organization, including peaceful organization, nonviolent reformers are driven underground and citizens in general may well become increasingly alienated. Indeed, they may gradually be transformed into active revolutionaries. The leader of the BFM, for example, is a former parliamentarian whom the regime has, in essence, made an outlaw. The problem is not limited to Bahrain. Saudi opposition groups use regime intransigence toward more-moderate figures to suggest that radical reform is the only solution possible.[1]

---

[1]See, for example, "Escalating the 'Case for Reform,'" CDLR Bulletin, 1995.

The Gulf states' preference for accommodation over confrontation could leave them vulnerable to foreign-inspired movements. Confronting Iran in the past proved manageable. Except for Bahrain, the Gulf states faced little threat from their Shi'a communities, which Iran claimed to lead. Should more-radical Sunni regimes take power, the Gulf regimes may have a far more difficult time cracking down on dissent while showing solidarity abroad. In addition, the decision to accommodate foreign regimes may come at the price of tolerating the spread of their subversive message.

Gulf regimes' use of co-optation, and the substantial safety net in general, strain national economies while contributing to overall levels of discontent. Because individuals who receive poor educations in nonpractical subjects such as religious studies receive lucrative government positions, there is little incentive or opportunity for them to train themselves for a modern economy. Similarly, the safety net decreases incentives for individuals to take entry-level jobs or to learn new skills.

THREATS TO THE U.S. PRESENCE

The primary problem posed by political violence in the Gulf today is political rather than military. So far, none of the groups or individuals active in the Gulf has shown signs of conducting sophisticated operations intended to disrupt U.S. military operations on behalf of foreign militaries. There is also little evidence that more-lethal future strategies of violence, such as information-disruption technologies or chemical and biological weapons, are part of the arsenal of groups or individuals in the Gulf. Unorganized violence, however, will remain a particular problem, and the current strategies used by Gulf governments are not only unable to stop it but may unintentionally foster it. Riots, occasional attacks on government or U.S. facilities, attacks on U.S. personnel, and other forms of political unrest are difficult to eliminate. The level of disaffection with the regimes is high enough (particularly in Saudi Arabia and Bahrain) that individuals will at times use violence.

Unorganized violence will not serve the immediate ends of a foreign power or cause operational problems for the United States, but it may destabilize the Gulf states and could contribute to political problems for a continued U.S. presence. The ideological flexibility many regimes employ to reduce anti-regime sentiment could lead to a drawdown in the U.S. presence. Many anti-regime forces oppose the U.S. military presence; therefore, to placate domestic opinion, it is possible that regimes will try to curtail, or at least play down, U.S. deployments. In the 1950s and early 1960s, Saudi Arabia yielded to pressure from Arab nationalists and reduced the U.S. economic and military presence. Islamists are making similar demands today.

Another concern is the response of the U.S. public. After terrorist attacks such as the Khobar bombing, many media and public officials sought to find fault with the U.S. military or political leadership rather than recognizing that casualties are an inherent danger in such a security environment. The consensus in the United States on the importance of the Gulf—a consensus recognized and shared by much of the U.S. Congress and media, and many people—is currently strong enough that occasional violence has yet to seriously undermine popular support for the U.S. deployment.

However, should this consensus weaken, future violence could lead to growing domestic pressure in the United States to reduce the U.S. presence or to avoid its deployment during times of heightened danger. To reduce this problem, U.S. leadership must make clear that casualties may occur despite the best of preparations.

But political violence is a fluid phenomenon, constantly shifting in response to countermeasures. Success in fighting violence in the Gulf may lead to political violence elsewhere in the world. Skilled radicals are particularly adept at finding chinks in defenses. Thus, the more effective force protection is in the Gulf, the more likely that some—though hardly most—radicals may strike outside the region in response. Only those radicals with access to substantial funds and organizational assistance will be able to engage in truly global terrorism. But the Gulf has long been home to such terrorists. In 1998, terrorists may have chosen to bomb U.S. embassies in Kenya and Tanzania because U.S. facilities in the Gulf are better fortified and on a higher state of alert, and because Gulf governments are more vigorous in their measures against radicals. This extra-regionality may be a trend of the future, with terrorists striking in Africa, Asia, or even the United States in response to events in the Gulf.[2]

EVALUATING SOLUTIONS FOR FIGHTING POLITICAL VIOLENCE

Preventing the above problems—and avoiding new ones—will be difficult. Several changes in U.S. military policy and diplomatic initiatives could, in theory, reduce political violence:

- Enacting political and economic reform in the Gulf

- Changing the U.S. presence in the region through new basing and operational approaches

- Increasing a European role in Gulf security

- Stopping foreign powers from contributing to violence

- Strengthening the U.S.–Gulf partnership

- Improving military-to-military ties.

But these changes come with their own set of trade-offs and will have only a limited impact. Each measure, along with its trade-offs and impact, is addressed below.

Enacting Political and Economic Reforms: The Mixed Benefits

If the Gulf regimes liberalized their economies and opened up decisionmaking, they could offset much of the hostility stemming from political alienation and economic discontent. A more open political system—and more genuine economic opportunities for those not tied to the ruling families—could decrease complaints stemming from corruption and political alienation.

The past record indicates, however, that the Gulf regimes have at best a limited recognition of the need for political and economic reform. Many Gulf leaders deliberately conflate anti-regime com-

---

[2]Hoffman, 1998.

plaints with support for violent radicals. The Al Khalifa, for example, tried to label all their opponents Iranian-backed terrorists, even though many of those involved in anti-regime protests had quite modest and even constructive agendas. Many Gulf rulers still regard their countries as private fiefdoms rather than as national lands for their citizens.

Even if they were willing, Gulf governments face severe constraints in their efforts to implement reform. A large and sustained rise in the price of oil is not expected, so regime revenues are not likely to increase dramatically in the coming years as local populations grow. Most regime officials are cautious and act only with a significant degree of elite consensus, making it hard for them to respond rapidly to new developments. Sweeping reforms are particularly difficult, because Gulf ruling families depend heavily on tradition to legitimate their rule. Furthermore, most political and economic reforms will directly affect the ruling family's own power and wealth, making it hard for rulers to gain support for such reforms among key decisionmakers, even when they recognize the need for change.

Moreover, a few government reforms will not easily rectify the many popular grievances stemming from deep socio-political and economic problems. Some complaints are particularly intractable, such as anger at Westernization or resentment over changing social mores. Nor are opposition agendas always realistic: Radicals often seek a total reorganization of society, a demand to which no government will accede.

In the short term, reform could conceivably exacerbate the "expectations gap." Any belt-tightening or even continued stagnation will only highlight that the government is not fulfilling its expected role of providing the good life. Moreover, opposition groups regularly play on the expectations gap, leaving any reforming regime vulnerable to their criticism. The regimes' lukewarm efforts to remove the social safety net and to make prices realistically reflect market levels have already engendered opposition criticism.[3]

Political reform in the Gulf and progress on human-rights issues are problematic for the United States. On the one hand, reform helps

---

[3]See "On Saudi Events," 1995.

the United States meet important goals related to human rights and the spread of democracy; on the other hand, it is a mixed blessing for stability. The development of democracy in any of these states could promote more-forceful expressions of anti-U.S. sentiment. Increased popular input into decisionmaking in Saudi Arabia would lead to greater pressure to reduce the U.S. presence in the Kingdom, because many among the population favor a decreased U.S. presence. Furthermore, peaceful democratization in one country could inspire activists in a neighboring state, which might lead to domestic unrest if the state responded negatively or with violence.

Human rights in general is a difficult subject. The repression that prevents anti–U.S. and anti-regime opposition from organizing also crushes liberal tendencies. Moreover, U.S. support for such rights can discredit reformers in states where anti–U.S. sentiment is high, such as Saudi Arabia. Likewise, U.S. efforts to foster such changes are likely to anger its allies in the region, making them less likely to cooperate on security matters.

U.S. support for measures to counter economic problems also will meet with difficulties. The myth of vast Gulf wealth is likely to make any foreign aid to Gulf governments politically impossible; indeed, the United States is likely to continue asking the Gulf states for economic assistance in supporting the U.S. military presence in the region. U.S. efforts to retrain Gulf workers, smooth the dislocations caused by privatization, or otherwise help speed economic reform are not likely to gain any takers in the Gulf, because the governments there remain hesitant to undertake serious reform.

Changing the U.S. Presence

The United States also might reduce the threat of political violence by altering its presence in the Gulf region. This could be done by decreasing its overall size, basing forces outside the immediate Gulf region, and boosting the ability to deploy rapidly with considerable force.

Reducing the Footprint. One option for reducing the threat terrorists pose to U.S. forces in the Gulf is to reduce the size of the U.S. "footprint"—the number of troops and visibility of the U.S. regional presence—there. Although the United States has reduced the visibility of its military personnel in recent years by moving them to more-isolated locations and restricting their time-off movement, the large numbers of troops make it impossible to diminish the impression of a considerable U.S. military force in the region. If the United States cut back its forces, particularly those stationed in Saudi Arabia, it would reduce the threat posed by terrorists and decrease the potential for political violence.

A smaller U.S. presence would reduce the targets available to terrorists. Currently, the large and diversified U.S. presence provides terrorist groups a variety of targets to strike. Moreover, the large numbers of U.S. troops offer "targets of opportunity" for less-capable but nevertheless violent groups or individuals. U.S. forces have offset these problems somewhat with remote basing and high levels of vigilance.

A reduction in the size of the U.S. footprint would also lessen some Gulf residents' complaints of foreign dependence, but it would not eliminate them completely. Some radicals oppose any U.S. presence in the Kingdom, seeing it as an indication of the regime's servility. Cutting the size of the presence will not placate these radicals. As one official noted, "the key is not the actual volume of the presence but the strength of the relationship. Size alone will not make a difference."[4]

The Iranian revolution is instructive. In Iran, the actual U.S. military presence was quite small, limited to military advisers and training missions. However, the Shah's relationship with Washington was close and open, discrediting him with many Iranian nationalists and Islamists.

A smaller footprint would also make a marginal contribution to improving the Gulf states' economy. Currently, several Gulf governments provide moderate amounts of support for the U.S. presence, paying for fuel and supplies and providing access to facilities—amounts that Gulf citizens nevertheless perceive to be quite large. Thus, a reduction would offset the criticism that the Gulf states are wasting their limited resources on the U.S. military.

---

[4]Authors' interview.

Finally, a smaller footprint may reduce incentives for outside states and movements to support terrorist acts in the Gulf region. Iran in particular has long railed against the U.S. military presence in the region and regularly calls for the United States to remove that presence. To the extent that Iranian-backed terrorism is motivated by fear of the U.S. military presence and an ideological commitment against U.S. regional hegemony, a drawdown in U.S. troops may reduce it. But a limited drawdown will not win the goodwill of more-radical figures, such as Osama bin-Ladin and his followers, who seek a complete withdrawal of U.S. forces from the region.

A smaller U.S. footprint carries a price: It might leave U.S. allies in the region more vulnerable to an attack from Iraq or Iran. Unfortunately, the GCC states are not ready to assume the burden for their own security. Military service is not respected among the GCC citizenry. Despite the relatively advanced weapon systems they own, the GCC militaries are generally of poor quality. Moreover, given the size disparities between the population of Gulf states and that of potential aggressors such as Iran or Iraq, the military imbalance between U.S. allies and these aggressors is likely to remain.

In view of these vulnerabilities, if the United States is to reduce its footprint in the region, it should continue its current approach of expanding the number of regional bases and integrating new technological and operational concepts to increase the lethality of its "over-the-horizon" presence—the proven ability of U.S. forces to deploy rapidly to the region to protect allies without actually being pre-deployed in their countries.

Finding New Basing. Perhaps the most promising option for minimizing the U.S. presence in vulnerable states such as Bahrain and Saudi Arabia is to find new base locations in the theater that will enable the United States to conduct operations effectively. As missile technologies spread and become more accurate, finding bases outside the Gulf will become necessary in any event.[5] To offset the

---

[5]Ensuring regional access will become more complicated in the coming years. The increasing availability of deep-strike weapons will make U.S. air bases more vulnerable to missile attacks. Several bases in Saudi Arabia are already vulnerable to missile strikes by Iran and Iraq. Should Iran develop and deploy the No-Dong missile, with its 1,300-kilometer range, basing sites throughout Southwest Asia will also be vulnerable. Thus, the U.S. military's tradition of operating air campaigns from bases immune to

threat to or loss of bases in the Gulf, the United States should mini-
mize its actual presence in the Gulf region while increasing
cooperation with friendly neighboring states, and should consider
returning to more of an over-the-horizon approach, particularly
since the U.S. presence often destabilizes the very allies it seeks to
protect. Ironically, the over-the-horizon presence on which the Gulf
states relied in the 1980s for their protection is far more feasible
today, even though the United States has expanded its ground
presence in the region.[6]

The United States should also make its regional posture more robust
by seeking to expand cooperation with other Middle Eastern countries, particularly Turkey. Turkey is a highly useful ally. During
Operation Desert Storm, the United States relied heavily on Turkish
bases, which are near Iran and Iraq (although they are less useful for
operations in the southern Gulf), to conduct air operations. Incirlik
and Diyabakir, for example, are 950 kilometers and 700 kilometers to
Baghdad, respectively, and 1,450 and 1,100 kilometers to Tehran, but
over 2,000 kilometers to the Strait of Hormuz. Bases in Turkey also
were valuable in various post–Desert Storm operations against Iraq.

However, security cooperation with Turkey is declining, for a variety
of reasons. With the end of the Cold War, U.S. military aid to Turkey
fell both in quality and quantity: eight of 12 NATO bases in Turkey
were shut down. Turkey also favors a balance-of-power approach in
the Gulf, seeking to play potential foes off against one another, rather
than show open hostility toward Iran and Iraq. It fears that the de
facto autonomy enjoyed by Iraqi Kurds under U.S. protection will
foster political instability among its own Kurdish population. Finally,
several Turkish leaders are concerned that Turkey's ties to the United
States contribute to Turkey's lack of influence in the Middle East.[7]

---

attack from enemy forces may be at an end. The vulnerability of bases will grow
tremendously should regional militaries use missiles to deliver nuclear, chemical, or
biological weapons. For an analysis of this problem and suggested approaches to
dealing with it, see Stillion and Orletsky, 1999.

[6]As well, there is no need for the military to conduct rest and recreation in the Gulf;
the extra cost and difficulty of flying troops to other regions are more than offset by the
decreased hostility and vulnerability that are likely to result from a lower level of
visibility.

[7]Aykan, 1996, pp. 346–352; Sariolghalam, 1996, pp. 304–309.

Despite these differences, the potential for closer relations is great. Turkey remains generally in sync with U.S. interests in the region. As does the United States, Turkey seeks stability in the former Soviet Union and the Persian Gulf. Both the United States and Turkey also are concerned about Syrian adventurism and support for terrorism. Ankara's feud with Syria has gone on for many years and includes issues such as water rights, Syrian support for Armenian and Kurdish terrorism, and border issues. Turkey also cooperates openly and frequently with Israel on security issues, despite the protests of Egypt, Syria, and Iran.

Bases outside the Gulf pose several problems for the U.S. military. Arranging such bases will require a major diplomatic effort that may entail making trade-offs in other areas of concern. Bases farther from the Gulf also will require a greater logistics effort and more assets: The forces' effectiveness will decline the farther they are from the theater. Perhaps most important, the deterrent effect of the U.S. presence may decrease: Regional aggressors are more likely to doubt that a brigade based outside the Gulf will play the same role as a brigade in Kuwait.

New Operational Approaches. In addition to new bases, increasing the speed and lethality of early-arriving forces will enable the United States to depend less on predeployed forces. The Expeditionary Air Force now being developed by the Air Force is one step in the right direction. The EAF will improve the Air Force's ability to deploy rapidly, making the U.S. presence more flexible and hence more useful for crises in Southwest Asia. Improving the lethality of early-arriving light infantry forces or more-rapid lift also would strengthen U.S. over-the-horizon capability. New operational concepts such as these improve the United States' ability to defeat regional aggression while minimizing its actual presence in the region during noncrisis periods.

New operational doctrines are less-than-perfect solutions, however. By definition, the effectiveness of such doctrines is not proven: In an actual crisis, unanticipated (by planner) problems with an EAF and other new approaches may develop, and the deterrent effect of the U.S. presence may be reduced by greater reliance on an over-the-horizon presence, even though this presence is lethal and effective. An over-the-horizon approach may encounter particular problems when access is not certain. If terrorists ever became sophisticated enough to act as special operations forces for invading armies, they might attack early-arriving forces and otherwise hinder the U.S. ability to deploy rapidly.

A Greater European Role

Another alternative for reducing the U.S. footprint is to increase the European presence in the region while decreasing the number of U.S. troops. European forces could play a greater role in helping U.S. forces administer the no-fly zone over Iraq. European battalions might also be incorporated into U.S. forces that conduct exercises in Kuwait and elsewhere in the region. Several European governments are considering reforms that would make their militaries better able to project power beyond Europe, thus making them more useful for Gulf contingencies.

A greater European role will reduce resentment of the United States, but it will hardly eliminate it. Considering it the standard-bearer of disruptive Western culture, many Islamists distrust the United States. However, a greater European presence will not completely offset hostility toward outside forces. Many Islamists complain of "Western" or of "infidel" forces—labels that presumably include the French and the British, as well as Americans. The replacement of U.S. forces with European troops also will not reduce popular resentment over the cost of the foreign presence. France and Britain have their own colonial legacy, which embitters many Arabs to this day.

Such a replacement, however, would have a greater impact on U.S. politics than on Gulf politics, perhaps increasing the chances that U.S. forces will deploy in times of crises. An increase in burdensharing—European forces will be in the line of fire, just as U.S. troops are today—will reduce criticism at home that the United States is bearing a disproportionate share of defending its allies. Furthermore, an allied presence in the region will add credibility to arguments that the region as a whole is vital to U.S. energy security.

A European presence will have mixed effects on the support of neighboring regimes for political violence. If European soldiers be-came a significant presence in the region, any indiscriminate terrorist attack would anger all Western states—not just Washington. However, Iran and Iraq will make few distinctions between U.S. and European troops if both are committed to the same mission: protecting the Gulf states from aggression by their neighbors. Moreover, although specific terrorism directed at the United States may diminish if the U.S. presence decreases, terrorism directed at British or French units may occur as a result of disputes between those countries and Gulf radicals that do not involve the United States. If casualties mount, this danger to non–U.S. forces could lead to recriminations and dissent between the United States and its allies.

U.S. allies, particularly Japan, Germany, and France, have shown little interest in increasing their role in providing for Gulf security, even though they too depend heavily on imported oil. These nations prefer to rely on Washington to act as the region's policeman. France has consistently pushed the United Nations to end sanctions on Iraq, backing down only after Iraq moves again to threaten a neighbor or UN inspectors unearth another weapons of mass destruction (WMD) cache. As the United States was trying to isolate Iran in 1996, Germany renewed its export-credit guarantees to Tehran. Similarly, France and Japan have rushed to trade with and invest in Iran. Indeed, the European powers do not even set markers for measuring the moderation they supposedly seek to achieve by engaging Iran.[8]

Even with the political desire to participate, European militaries currently lack the power-projection assets and rapidly deployable forces necessary to play a major role in the Persian Gulf. Western European militaries do not have organized military sealift. It would take several weeks or more for heavy Western European assets to arrive in areas as far away as the Persian Gulf, even if the United States helped. Western European militaries also currently lack projectable command, control, communication, and intelligence (C4I) assets and logistics support for operations outside Europe, and their navies are designed for coastal defense, not blue-water power projection. Furthermore, many European militaries maintain a large portion of their forces on reserve status and so are not able to deploy on short

---

[8]Garfinkle, 1997, p. 24.

notice.[9] Several European militaries are considering changes to their force posture that will make them more suited to operations in the Gulf—though how well suited is too soon to tell. This development is promising, and the United States should encourage it.

Because of these many problems, any European contribution to Gulf security is likely to be limited at best in the near term. Even the nominal augmentation of the U.S. presence with European forces will foster coordination difficulties, which will be compounded by the differential in readiness and training between the United States and its allies. In addition, the United States will have to make political concessions in the Gulf to keep its European allies satisfied. This may require a softer U.S. line toward Iran or Iraq, even if neither regime moderates its behavior.

Halting Foreign Interference

In addition to relying on external powers for defense, the Gulf states could also try to reduce the ability of Iran, Iraq, and other foes to stir up internal unrest. The Gulf states have successfully fought repeated attempts by foreign governments to destabilize their countries, but they will find it difficult to end such interference completely. Most important, events outside the Gulf can, and will, inspire Gulf youths. Images of brave Palestinian *fedayeen* or zealous Afghan *mujahedin* serve as role models for Gulf youths, leading them to see violence as an acceptable form of political action and an inevitable stimulus to political change. Similarly, the intellectual environment in the Gulf is influenced by radical thinkers from Egypt and elsewhere. These intellectuals and theologians often provide an ideological justification for violent action, even when local intellectuals support their regimes. Thus, anti-Saudi activists make a point of declaring the Al Saud un-Islamic, an attack that compels the faithful to resist their ruling family. Altering the local intellectual environment is almost impossible: Long a part of the larger Middle East community, the Gulf is tied to the broader Arab and Muslim community.

---

[9]Kugler, 1994, p. 80. However, France and Britain have multidivision permanent standing forces that they could contribute.

Completely stopping direct foreign intervention also will be difficult, if not impossible. On their own, the Gulf states lack the military means to threaten foreign governments directly. Instead, they have used an accommodative foreign policy to try to gain their neighbors' and other radicals' goodwill. Yet, despite such a policy, both Iran and Iraq could seek to incite unrest in the Gulf for strategic, domestic, or ideological reasons, particularly if they saw it as a tool for weakening the U.S. presence on the peninsula. Such support could involve infiltrating provocateurs into the Gulf, and training local radicals or providing funds to them to recruit members and buy weapons. In addition, foreign governments could issue calls for rebellion and violence, which might inspire sympathetic domestic groups. Over time, foreign agents and sympathetic activists are likely to be identified and arrested, but short-term violence remains possible.

In theory, if Iran or Iraq sponsored political violence in the Gulf, the United States could assist its allies by using military force against either or both countries. By threatening to use force, the United States could try to compel state sponsors to halt support for political violence. Thus, such threats might reduce foreign assistance to terrorist groups in the Gulf. The threat of punishment has sometimes influenced Iranian and Iraqi behavior.[10] For example, Iran tied the hands of many of its forces after Desert Storm—in response to a U.S. warning not to intervene—even though Iraqi Shi'a were being butchered.

Nevertheless, how military retaliation affects political violence is difficult to judge.[11] U.S. threats against Iran and Iraq will make both adversaries hesitate before challenging the United States and its allies. But the past record indicates that they often engage in terrorism despite U.S. threats and warnings.

---

[10]Attacks on state sponsors also have an indirect effect: They convince U.S. allies of the serious nature of U.S. policy. After the 1986 El Dorado Canyon mission against Libya, both friends and enemies abroad paid more attention to U.S. statements on terrorism. Indeed, some sponsors urged their clients not to use terrorism for fear of a U.S. military response. Tucker, 1997, pp. 39–40.

[11]Studies have demonstrated no direct correlation between a military response and a fall in terrorist incidents. However, such studies are limited by the difficulty of proving a counterfactual: At what level would attacks have been if the retaliation had not occurred? See Miller, 1990, p. 120, for an assessment.

The model for deterring terrorists through direct attacks that many observers have in mind when discussing military strikes is the tactics Israel used against Palestinian political violence. Israel deployed military force against terrorist personnel, bases, and facilities. The intention was to inflict unacceptable damage on the terrorists and to deter host nations from providing the terrorists safe haven. As such, Israel struck its neighbors, hoping to discourage support for and toleration of the PLO. In Jordan, this policy helped to precipitate a government crackdown against Palestinian radicals in 1970. In Lebanon, however, this policy backfired and contributed to the descent of Lebanon into civil war—chaos that gave terrorists even more autonomy. Israeli strikes in Lebanon also created a new problem—Shi'a political violence—that soon dwarfed the original threat posed by the Palestinians.

The Israeli model is difficult to apply to the Gulf, for several reasons. First, the United States would probably engage in such retaliation with limited support from the Gulf populaces or from the American people. Thus, retaliation could lead to greater resentment of the United States in key host nations. Second, the Israeli model requires a sustained campaign. Although Israel successfully coerced Jordan to crack down on Palestinian radicals, the coercion process took years. Expecting a single strike to accomplish the process is a mistake.

Moreover, both Iran and Iraq can be difficult to coerce. Both nations have engaged in behavior opposed by the United States (seeking weapons of mass destruction, engaging in political violence, etc.) despite U.S. sanctions and even limited military clashes. U.S. allies are reluctant to join in comprehensive sanctions on Iran, and their support for sanctions against Iraq is wearing thin. In Iran, few leaders would risk the charge of bowing to U.S. pressure.[12]

Effectively targeting nonstate sponsors of political violence—such as the Lebanese Hezbollah—is even more difficult. Such organizations are diffuse and lack assets that can be threatened. Moreover, identifying the members of these organizations is difficult. As well, Hezbollah enjoys the protection of Syria, and strikes against the

---

[12]For more on coercing Iraq, see Byman, Pollack, and Waxman, 1998.

movement would worsen regional relations and further strain the already-moribund peace process.

Strengthening the Gulf–U.S. Partnership

Decreasing the threat of political violence requires strong U.S.–Gulf cooperation, which will improve intelligence sharing and enable the United States and its allies to present a united front against Iran, Iraq, or other regional threats. At the same time, better cooperation may require changes in the way that the United States approaches the problem of terrorism in the Gulf—changes that may prove impossible for political or legal reasons.

The U.S. political system often hinders cooperation. The media leaks that are part and parcel of the U.S. political debate often enrage Gulf allies, who are embarrassed by accusations or revelations of their corruption or stonewalling. Publicity from any counterterrorism successes, which U.S. officials may at times seek, must be minimized in order to avoid straining relations with countries in which there is little popular support for a significant U.S. role.

The U.S. penchant for legalism also strains relations: Regional allies do not share U.S. concerns for a high burden of proof, the rule of law over political expediency, and legal accountability. Sharing more information with Gulf security services poses several legal and political problems for the United States, particularly if these states use that information to repress legitimate political opposition at home. Moreover, U.S. legal standards require a higher burden of evidence—and often the public disclosure of sensitive information— than does Saudi Arabia, which may seek the extradition of individuals protected by U.S. courts.

Saudi leaders and other Gulf state residents also may be hesitant to share information simply because they do not want the perpetrators caught. When terrorists—including even the notorious Osama bin-Ladin, who is from an important Saudi family—are linked to the ruling family or other Kingdom notables, the Saudis would prefer that the problem be handled quietly and with no outside involvement. If foreign sponsors are involved, Riyadh may try to avoid recriminations that could jeopardize political ties, particularly when relations are improving.

The United States and its allies often disagree on what is the proper policy to pursue toward Iran or Iraq, making it difficult to sustain a strong coercive campaign against either. The Gulf states often try to hedge their bets, maintaining a strong but low-key security arrangement with Washington while making friendly overtures toward Tehran and Baghdad. A U.S. military strike or other strong response would upset this delicate balance.

Improving Military-to-Military Ties

The United States and the Gulf can also improve cooperation in the military realm. As Gulf military officers come to know their U.S. counterparts better, military professionalism could unite the militaries, despite cultural and political differences. Better military-to-military ties would improve U.S. intelligence capabilities and foster goodwill toward the United States. Better ties might reduce the xenophobia common to Gulf societies, particularly Saudi Arabia. In addition, U.S. contacts with military officers will improve U.S. intelligence capabilities, providing another window on Gulf society.

However, better ties will contribute only modestly toward reducing political violence. The vast majority of the people likely to be hostile to the United States, particularly Islamists, are not likely to be touched at all by military-to-military ties, and they may even see closer ties as proof of the regime's corruption. In many Gulf states, the military's composition is limited to certain tribes or strata of society—groups that are often already loyal to the regime and thus need relatively little additional attention—thus limiting the exposure of much of the population to the goodwill generated by military-to-military ties. Indeed, in several states, much of the armed forces is staffed by foreigners, whose loyalty is not in doubt.

FINAL WORDS

The sources of political violence are simply too strong in the Gulf to end the threat completely. The various measures described above, particularly when employed in combination, can reduce the problem of political violence. But decisionmakers must recognize that several of these measures are difficult to implement. In addition, many measures have trade-offs that will affect other U.S. regional goals or will require additional sacrifices on the part of the U.S. military.

When trying to anticipate the future of violence in the Gulf, Washington must recognize that neither reform nor the status quo offers a perfect way out. There is no way to square this circle: Reform carries with it severe problems, some of which may make the U.S. presence in the Gulf still harder to sustain. And efforts to decrease the U.S. presence will at best reduce, but not eliminate, dangers. Thus, the need for robust force protection continues even if progress is made in reducing the causes of political violence.

Ajami, Fouad, *The Arab Predicament*, New York: Cambridge University Press, 1982 (revised 1991).

———, *The Dream Palace of the Arabs: A Generation's Odyssey*, New York: Pantheon Books, 1998.

Amos, John W., II, "Terrorism in the Middle East: The Diffusion of Violence," in Yonah Alexander, ed., *Middle East Terrorism: Current Threats and Future Prospects*, New York: G. K. Hall & Co., 1994.

Anderson, Lisa, "Absolutism and the Resilience of Monarchy in the Middle East," *Political Science Quarterly*, Vol. 106, No. 1, 1991, pp. 1–15.

Aykan, Mahmut Bali, "Turkish Perspectives on Turkish–U.S. Relations Concerning Persian Gulf Security in the Post–Cold War Era: 1989–1995," *Middle East Journal*, Vol. 50, No. 3, Summer 1996, pp. 344–358.

"Bahrain: Alleged Conspiracy Used As a Cover for Consolidating Tribal Dictatorship," Bahrain Freedom Movement communiqué, June 3, 1996.

"Bahrain: Defendants' Confessions Reported," Manama WAKH [Bahraini radio service], FBIS-NES-96-110, June 5, 1996.

"Bahrain: Economy Goes down As Al Khalifa Imports More Foreign Troops," Bahrain Freedom Movement communiqué, February 20, 1997.

"Bahrain: Interior Ministry on Arrest of 'Hizballah of Bahrain' Group," Manama WAKH [Bahraini radio service], FBIS-NES-96-107, June 3, 1996.

"Bahrain Uprising: 3 Years Old," Bahrain Freedom Movement e-mail, December 4, 1997.

Bahry, Louay, "The Opposition in Bahrain: A Bellwether for the Gulf?" *Middle East Policy*, Vol. 5, No. 2, May 1997, pp. 42–57.

Bodansky, Yossef, "Iranian and Bosnian Leaders Embark on a New Major Escalation of Terrorism Against the West," *Defense and Foreign Affairs Strategic Policy*, Vol 21, No. 8, August 31, 1993 (electronic version).

Brush, Stephen G., "Dynamics of Theory Change in the Social Sciences: Relative Deprivation and Collective Violence," *Journal of Conflict Resolution*, Vol. 40, No. 4, 1996, pp. 523–545.

Byman, Daniel, Kenneth Pollack, and Matthew Waxman, "Coercing Saddam Hussein: Lessons from the Past," *Survival*, Autumn 1998.

CIA Handbook of International Economic Statistics, 1996, Washington, DC: U.S. Government Printing Office, 1996.

CIA World Factbook, 1996, electronic version, accessed on November 18, 1997 (http://www.odci.gov/cia/publications/factbook/country-frame.html).

Cordesman, Anthony, *Bahrain, Oman, Qatar, and the UAE*, Boulder, Colorado: Westview Press, 1997a.

———, *Saudi Arabia: Guarding the Desert Kingdom*, Boulder, Colorado: Westview Press, 1997b.

Crystal, Jill, *Oil and Politics in the Gulf: Rulers and Merchants in Kuwait and Qatar*, New York: Cambridge University Press, 1995.

Dabashi, Hamid, *Theology of Discontent: The Ideological Foundations of the Islamic Revolution in Iran*, New York: New York University Press, 1993.

Della Porta, Donatella, "Left Wing Terrorism in Italy," in Martha Crenshaw, ed., *Terrorism in Context*, University Park, Pennsylvania: The Pennsylvania State University Press, 1995, pp. 105–159.

Demographic Yearbook 1986, New York: United Nations, 1988.

Dunn, Michael Collins, "Five Years After Desert Storm: Security, Stability, and the U.S. Presence," *Middle East Policy*, Vol. 4, No. 3, 1996.

————, "Is the Sky Falling? Saudi Arabia's Economic Problems and Political Stability," *Middle East Policy*, Vol. III, No. 4, April 1995, pp. 29–39.

"Escalating the 'Case for Reform,'" Bulletin No. 37 of the Committee for the Defense of Legitimate Rights of Saudi Arabia, FBIS-NES-95-051, March 3, 1995 (fax).

Fakhro, Munira A., *Economy, Security, and Religion*, New York: St. Martin's Press, 1997, pp. 167–188.

Fandy, Mamoun, "From Confrontation to Creative Resistance: The Shia's Oppositional Discourse in Saudi Arabia," *Critique*, Fall 1996, pp. 1–27.

Garfinkle, Adam, "The U.S. Imperial Postulate in the Mideast," *Orbis*, Winter 1997.

Gause, Gregory F., *Oil Monarchies: Domestic and Security Challenges in the Arab Gulf States*, New York: Council on Foreign Relations, 1994.

————, "The Political Economy of National Security in the GCC States," in "The Coming Crisis in the Persian Gulf," Chapter 3 in Gary Sick and Lawrence G. Potter, eds., *The Persian Gulf at the Millennium: Essays in Politics, Economy, Security, and Religion*, New York: St. Martin's Press, 1997, pp. 61–84.

Ghabra, Shafeeq, "The Islamic Movement in Kuwait," *Middle East Policy*, Vol. V, No. 2, May 1997, pp. 58–72.

————, "Voluntary Associations in Kuwait: The Foundation of a New System?" *Middle East Journal*, Vol. 45, No. 2, Spring 1991, pp. 199–215.

Goodwin, Jeff, and Theda Skocpol, "Explaining Revolutions in the Contemporary Third World," *Politics and Society*, Vol. 17, December 1989.

Green, Jerrold D., *Revolution in Iran: The Politics of Countermobilization*, New York: Praeger, 1982.

————, "Terrorism and Politics in Iran," in Martha Crenshaw, ed., *Terrorism in Context*, University Park, Pennsylvania: The Pennsylvania State University Press, 1995, pp. 553–594.

Gurr, Ted Robert, *Minorities at Risk: A Global View of Ethnopolitical Conflicts*, Washington, D.C.: United States Institute of Peace, 1993.

Hedges, Chris, "Foreign Islamic Fighters in Bosnia Pose a Potential Threat for G.I.'s," *The New York Times*, December 3, 1995, p. 1.

Henderson, Simon, *After King Fahd: Succession in Saudi Arabia*, Washington, D.C.: The Washington Institute, Policy Paper 37, 1994.

Hoffman, Bruce, *Inside Terrorism*, New York: Columbia University Press, 1998.

————, *Recent Trends and Future Prospects of Iranian-Sponsored International Terrorism*, Santa Monica, Calif.: RAND, R-3783-USDP, 1990.

————, "The New Terrorist: Mute, Unnamed, Bloodthirsty," *The Los Angeles Times*, August 16, 1998.

Jaber, Hala, *Hezbollah: Born with a Vengeance*, New York: Columbia University Press, 1997.

al-Jamri, Mansoor, "Prospects of a Moderate Islamist Discourse: The Case of Bahrain," Paper given at the November 22, 1997, Middle East Studies Association Meeting in San Francisco, California.

Kellen, Konrad, *Terrorists—What Are They Like? How Some Terrorists Describe Their World and Actions*, Santa Monica, Calif.: RAND, N-1300-SL, 1979.

Khalilzad, Zalmay, David Shlapak, and Daniel Byman, *The Implications of the Possible End of the Arab-Israeli Conflict for Gulf Security*, Santa Monica, Calif.: RAND, MR-822-AF, 1997.

al-Khoei, Youssef, "The Shi'a of Medina," *Dialogue*, July 1996 (downloaded from the Gulf 2000 database).

Kugler, Richard L., *U.S.–West European Cooperation in Out-of-Area Military Operations: Problems and Prospects*, Santa Monica, Calif.: RAND, MR-349-USDP, 1994.

"Kuwait: Youths Said Seizing Satellite Dishes, VCRs," Cairo Al-Akhbar [Egyptian newspaper], FBIS-NES-97-024, January 30, 1997, p. 2.

Landau, Jacob M., *The Politics of Pan-Islam*, Oxford: Clarendon Press, 1994.

Miller, Reuben, "Responding to Terrorism's Challenge: The Case of Israeli Reprisals," *Virginia Social Science Journal*, Vol. 25, 1990, pp. 109–123.

Mitchell, Richard P., *The Society of Muslim Brothers*, New York: Oxford University Press, 1969 (revised 1993).

Momen, Moojan, *An Introduction to Shi'a Islam*, New Haven: Yale University Press, 1985.

Mottahedeh, Roy P., and Mamoun Fandy, "The Islamic Movement: The Case for Democratic Inclusion," in "The Coming Crisis in the Persian Gulf," Chapter 11 in Gary Sick and Lawrence G. Potter, eds., *The Persian Gulf at the Millennium: Essays in Politics, Economy, Security, and Religion*, New York: St. Martin's Press, 1997, pp. 297–318.

"The Movement for Islamic Reform in Arabia," July 14, 1998, communiqué, e-mail version.

Muller, Edward N., *Aggressive Political Participation*, Princeton: Princeton University Press, 1979.

Muller, Edward N., and Thomas O. Jukam, "Discontent and Aggressive Political Behavior," *British Journal of Political Science*, 1983.

O'Ballance, Edgar, *Islamic Fundamentalist Terrorism, 1979–1995*, New York: New York University Press, 1997.

"On Saudi Events," CDLR Bulletin No. 31, FBIS-NES-95-027, January 20, 1995.

Ranstorp, Magnus, *Hizb'Allah in Lebanon: The Politics of the Western Hostage Crisis*, New York: St. Martin's Press, 1997.

Rugh, William A., "What Are the Sources of UAE Stability?" *Middle East Policy*, Vol. V, No. 3, September 1997, pp. 14–24.

"Saddam's Spies," *Gulf States Newsletter*, November 6, 1995.

Sariolghalam, Mahmood, "The Future of the Middle East: The Impact of the Northern Tier," *Security Dialogue*, Vol. 27, No. 3, 1996, pp. 303–317.

"Saudi Arabia: Correspondent Meets with Opposition Leader Bin-Ladin," London Channel 4 Television Network, FBIS-NES-97-035, February 20, 1997.

al-Shayeji, Abdullah, "Gulf Views of U.S. Policy in the Region," *Middle East Policy*, Vol. 5, No. 3, September 1997.

Sick, Gary G., "The Coming Crisis in the Persian Gulf," Chapter 1 in Gary Sick and Lawrence G. Potter, eds., *The Persian Gulf at the Millennium: Essays in Politics, Economy, Security, and Religion*, New York: St. Martin's Press, 1997, pp. 13–30.

Skocpol, Theda, *States and Social Revolutions*, Cambridge: Cambridge University Press, 1979.

Statistics Yearbook, 1976, New York: United Nations, 1978.

Statistics Yearbook, 1983/1984, New York: United Nations, 1985.

Statistics Yearbook, 1993, New York: United Nations, 1995.

Stillion, John, and David T. Orletsky, *Airbase Vulnerability to Conventional Cruise-Missile and Ballistic-Missile Attacks: Technology, Scenarios, and U.S. Air Force Responses*, Santa Monica, Calif.: RAND, MR-1028-AF, 1999.

Tilly, Charles, *From Mobilization to Revolution*, Reading, Mass.: Addison-Wesley, 1978.

Tucker, David, *Skirmishes at the Edge of Empire: The United States and International Terrorism*, Westport, Conn.: Praeger, 1997.

Wardlaw, Grant, *Political Terrorism: Theory, Tactics, and Counter-measures*, New York: Cambridge University Press, 1982 (revised 1990).

"Your Right to Know: The End of the Deference Era," CDLR Bulletin No. 30, FBIS-NES-95-027, Bulletin released on January 13, 1995 (Internet version).

Zonis, Martin, *The Political Elite of Iran*, Princeton: Princeton University Press, 1971.